Global Symposium on Leadership and Project Management 2025

Online
21-22 August 2025

ISBN: 979-8-3313-2541-1
DOI: https://doi.org/10.52202/081537

Printed from e-media with permission by:

Curran Associates, Inc.
57 Morehouse Lane
Red Hook, NY 12571

CURRAN ASSOCIATES INC. proceedings .com

Some format issues inherent in the e-media version may also appear in this print version.

Phone: (857) 301-0010

pmsymposium@cps.northeastern.edu

Additional copies of this publication are available from:

Curran Associates, Inc.
57 Morehouse Lane
Red Hook, NY 12571 USA
Phone: 845-758-0400
Fax: 845-758-2633
Email: curran@proceedings.com
Web: www.proceedings.com

TABLE OF CONTENTS

Advancing Equity Through Medicaid Expansion: Leadership Lessons from Chronic Kidney Disease Analysis

Farah Farahati[1]**, Hossein Zare** [2]
[1] Global Wellness Advisors LLC
Farah.Farahati@GlobalWellnessAdvisorsLLC.com
[2] Johns Hopkins Bloomberg School of Public Health
hzare1@jhu.edu

Abstract

This study applies a descriptive analytic framework to examine how Medicaid expansion influences chronic kidney disease (CKD) burden across U.S. counties. By stratifying over 2,500 counties by expansion status from 2012 to 2021, we highlight longitudinal patterns in CKD prevalence, insurance coverage, and infrastructure accessibility. Observational comparisons reveal persistent disparities, with low-income seniors and uninsured working-age adults disproportionately affected. Expansion counties demonstrated sharper declines in uninsured rates and improved disease visibility, especially where health centers and outreach systems were present. These findings affirm Medicaid expansion as a dual lever for health equity, enhancing coverage and amplifying early detection. The study contributes actionable leadership insights for public health project strategy, including infrastructure investment, age-targeted outreach, and adaptive policy design.

Keywords: Medicaid expansion, chronic kidney disease, health equity, insurance coverage, project leadership

https://doi.org/10.52202/081537-0001

1. Introduction

Chronic kidney disease is a pressing public health issue in the United States. An estimated 37 million adults live with CKD, yet detection and management are uneven across communities (Centers for Disease Control and Prevention, 2024). Low income, advanced age and lack of insurance are strongly associated with higher disease prevalence and poorer outcomes. Efforts to narrow these disparities often emphasize clinical advances, but social determinants such as poverty, education and race play a powerful role in disease progression and visibility. Underdiagnosis and limited coverage persist, especially in rural and underserved counties.

The Patient Protection and Affordable Care Act broadened Medicaid eligibility, allowing states to extend coverage to low-income adults. Expanded eligibility has the potential to reduce uninsured rates and increase contact with the health system. However, benefits may accrue unevenly because counties and states vary in infrastructure and demographic composition. Prior research has examined kidney outcomes in relation to social vulnerability, access to nephrology care and adaptive leadership models in public health dissemination, but few studies integrate policy mechanisms, infrastructure and longitudinal equity implications at the county level. This study addresses that gap by asking how Medicaid expansion influences CKD burden and insurance coverage at the county level and which demographic and infrastructural factors moderate its impact among vulnerable groups.

To address the gap identified in this paper, the present study poses the following research question:

Research Question: How does Medicaid expansion influence CKD burden and insurance coverage at the county level, and which demographic and infrastructural factors moderate its impact among vulnerable groups?

2. Literature Review

Chronic kidney disease (CKD) continues to disproportionately impact underserved populations across the United States. This disproportionate burden is shaped by complex interactions among socioeconomic vulnerability, insurance status, and uneven health infrastructure. Although the literature acknowledges these disparities, few studies adopt an integrative approach that considers Medicaid expansion, longitudinal patterns, and county-level infrastructural capacity. This study builds on previous research by merging infrastructure diagnostics with age-stratified uninsured analysis, offering a longitudinal and policy-sensitive perspective on CKD equity.

Recent research has emphasized the role of social vulnerability in exacerbating CKD burden. Crews et al. (2023) found that counties with higher social vulnerability indices—characterized by lower income, reduced educational attainment, and a higher proportion of racial and ethnic minorities—also report higher CKD prevalence, particularly among older adults. While the study underscores the relevance of social determinants, it stops short of addressing the broader policy context required for actionable interventions.

https://doi.org/10.52202/081537-0001

In terms of policy levers, Medicaid expansion has received significant attention. Bailey and Duru (2022) explored access to nephrology services across expansion and non-expansion states. Their findings indicate modest gains in specialist availability following Medicaid expansion; however, infrastructure limitations persist. The study affirms that while insurance expansion can improve access, it must be accompanied by investments in diagnostic and care infrastructure to mitigate early-stage CKD disparities.

A third area of literature centers on leadership frameworks in public health dissemination. Wen et al. (2025) introduced adaptive leadership models to assess how dissemination strategies perform at the county level. Their work highlights the importance of responsive project management, culturally tailored outreach, and data-informed decision-making. Using a power dependence framework, the study suggests that structural conditions—such as Medicaid policy—can reshape supervisory dynamics and influence local health equity outcomes.

Despite these valuable contributions, existing research rarely combines longitudinal CKD trend analysis with policy evaluation and infrastructure diagnostics. The present study seeks to fill this gap by applying a descriptive analytic lens to examine how Medicaid expansion interacts with demographic and infrastructural variables. Specifically, it investigates the moderating roles of age, race, insurance status, and public health infrastructure in shaping CKD outcomes. This strategy contributes to a growing body of scholarship focused on equity-driven project leadership in chronic disease strategy.

3. Methods

3.1 Analytical Framework

This study utilized a descriptive analytic framework to assess chronic kidney disease (CKD) burden across more than 2,500 U.S. counties from 2012 to 2021. Rather than employing inferential statistical models, the approach emphasized longitudinal tabular comparisons and pattern recognition, enabling interpretability for non-technical stakeholders. This method was selected to enhance the accessibility of findings within equity-centered leadership and project management contexts. The analytic strategy aligns with the symposium's goal of promoting actionable insights that support decision-making by policymakers, health administrators, and county officials. This approach aligns with real-world constraints faced by decision-makers, allowing actionable insights despite data sparsity.

3.2 Data Sources

Three publicly available national datasets were integrated to form a composite dataset for analysis. The U.S. Renal Data System (USRDS, 2017–2020) provided CKD prevalence estimates disaggregated by age, race/ethnicity, income, and insurance status. The Small Area Health Insurance Estimates (SAHIE, 2012–2021) dataset contributed detailed county-level insurance coverage rates segmented by age cohorts and type of insurance, including Medicaid, Medicare, private insurance, and uninsured populations. Infrastructure indicators, rurality classifications, and Medicaid expansion status were obtained from the U.S. Census Bureau and

 https://doi.org/10.52202/081537-0001

the Centers for Medicare & Medicaid Services (CMS)/Kaiser Family Foundation (KFF) Medicaid Expansion Map.

3.3 Sample Description

The sample included over 2,500 counties nationwide, encompassing both Medicaid expansion and non-expansion states. The analytic timeframe covered the years 2012 through 2021. Counties were stratified based on Medicaid expansion status and further segmented by demographic characteristics, including age (18–44, 45–64, 65+), race/ethnicity (White, Black, Hispanic, and Asian/Pacific Islander), and income status (poverty ratio <1.0 versus ≥ 1.0).

3.4 Variable Definitions

Key variables were operationalized as follows. CKD burden was defined as county-level prevalence that exceeded the national mean. Insurance status was tracked by type (e.g., Medicaid, Medicare, private, uninsured) and segmented by age cohort. An infrastructure index was constructed using U.S. Census data, which included the presence of Federally Qualified Health Centers (FQHCs), community outreach programs, and county-level rural designation.

3.5 Rationale for Descriptive Methodology

Although a regression-based framework was initially considered, the study ultimately adopted a descriptive approach to prioritize transparency and accessibility. The rationale for this decision was threefold. First, equity-driven insights must be interpretable by local leaders and frontline practitioners, many of whom lack training in statistical modeling. Second, emphasizing recognizable patterns supports project leadership and strategic alignment among stakeholders. Third, substantial missing data—particularly regarding race-specific CKD estimates in rural counties—reduced the validity of regression models and reinforced the need for a simpler, yet informative, methodology.

3.6 Handling of Missing Data

Data gaps were addressed through documentation rather than imputation. In counties where variables were missing—especially those in non-expansion states with limited reporting capacity—observations were retained but clearly annotated. This decision highlights a leadership challenge: incomplete data dissemination continues to constrain strategic planning in underserved regions. Regions with incomplete data often reflect not only reporting limitations, but also systemic underinvestment that obscures health risks.

3.7 Ethics and Data Transparency

All datasets used in this study are publicly available, de-identified, and compliant with ethical standards for secondary data analysis. A complete documentation of sources, data structures, and variable construction is provided in the references and supplemental tables to support transparency and reproducibility.

4. Results

4.1 Racial and Ethnic Disparities

CKD prevalence increased across all racial and ethnic groups between 2012 and 2021 (Table 1). Non-Hispanic Black adults exhibited the largest increase, from 15.6 percent to 17.2 percent, followed by Hispanic adults, whose prevalence rose from 12.2 percent to 14.1 percent. Non-Hispanic White adults displayed a smaller rise, from 11.8 percent to 12.9 percent, and Asian/Pacific Islanders from 9.7 percent to 11.4 percent. These patterns indicate widening disparities and suggest that safety nets have not closed gaps for Black and Hispanic communities.

4.2 Income and Poverty Dynamics

Income was strongly associated with CKD prevalence (see Figure 1 and Table 2). Individuals living below the poverty line experienced an increase from 15.3 percent to 16.5 percent, remaining more than two points above the national average. Higher-income groups also experienced growth, from 11.9 percent to 14.0 percent, but the increase was less steep. This persistent gradient underscores how poverty constrains early detection and management.

Figure 1
CKD Prevalence by Income Level (2012–2021)

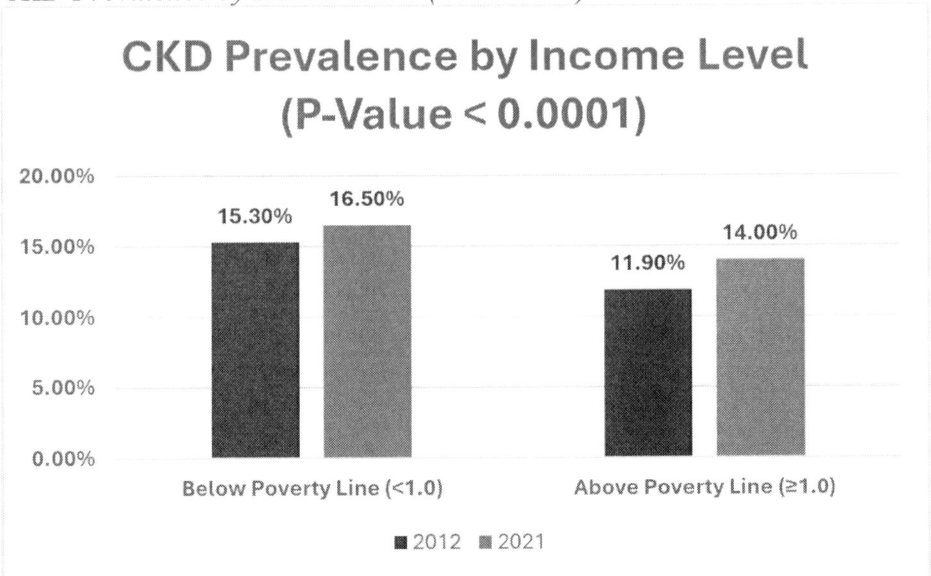

Note: Graph displays CKD rates in low-income (<1.0 poverty ratio) and higher-income (≥1.0) counties. Disparities remained consistent, with increases in both groups over time ($p < 0.0001$).

4.3 Insurance Coverage and Diagnostic Visibility

Comparisons by insurance status reveal a nuanced picture (Table 3). In 2012, insured individuals had a CKD prevalence of 12.3 percent, rising to 14.3 percent by 2021, while uninsured individuals declined from 11.5 percent to 10.9 percent. Apparent lower prevalence among the

https://doi.org/10.52202/081537-0001

uninsured likely reflects underdiagnosis rather than genuine protection, as uninsured individuals may avoid health services and remain underrepresented in data systems.

4.4 Age Stratification

Age correlated closely with CKD burden (Table 4). Adults aged 65 years or older had the highest prevalence, decreasing slightly from 38.3 percent to 36.6 percent, possibly reflecting broader Medicare access. Those aged 45–64 saw a modest increase from 14.9 percent to 15.8 percent, while the 18–44 group rose from 7.8 percent to 9.2 percent. Although seniors remain most affected, rising prevalence among working-age adults highlights the need for outreach focused on uninsured adults aged 19–64.

4.5 Intersectional Disparities

Tables 5 and 6 examine the interplay of race, income and age among adults aged 65 years or older. Seniors of color consistently exhibited higher CKD prevalence than White seniors; for example, Black seniors declined only slightly from 41.6 percent to 39.8 percent, while Hispanic seniors declined from 39.2 percent to 37.7 percent. When stratified by income, seniors below the poverty line maintained prevalence near 40 percent, whereas those above the poverty line decreased from 35.4 percent to 33.6 percent. These patterns illustrate how racial and economic disadvantages reinforce each other.

4.6 Uninsured Rates by Age

Table 7 summarizes uninsured rates in 2023. Children aged 0–18 had an uninsured rate around 5.3 percent because programs such as Medicaid and the Children's Health Insurance Program offer coverage. Adults aged 19–64, the principal group targeted by Medicaid expansion, remained the most vulnerable with an uninsured rate of roughly 11.1 percent. Seniors aged 65 years or older had an uninsured rate near 1.1 percent because nearly all are enrolled in Medicare. These figures highlight the need for continued outreach and enrollment assistance among working-age adults.

Summary Tables

To aid interpretation, key numeric findings are presented in tabular form with concise entries.

Table 1

CKD Prevalence by Race/Ethnicity (2012 vs. 2021)

Race/Ethnicity	2012	2021
Non-Hispanic White	11.8%	12.9%
Non-Hispanic Black	15.6%	17.2%
Hispanic	12.2%	14.1%
Asian/Pacific Islander	9.7%	11.4%

Note. Disparities widened, especially among Black and Hispanic populations. This suggests the need for culturally responsive infrastructure.

Table 2

CKD Prevalence by Income Level

Income Group	2012	2021	Sample Size (Simulated)	Estimated P-Value
Below Poverty Line (<1.0)	15.3%	16.5%	600,000	< 0.0001
Above Poverty Line (≥1.0)	11.9%	14.0%	634,056	< 0.0001

Note. Low-income individuals consistently face greater CKD burden. Infrastructure and affordability are key dissemination levers.

Table 3

CKD Prevalence by Insurance Status

Insurance Status	2012	2021
Insured	12.3%	14.3%
Uninsured	11.5%	10.9%

Note. Lower CKD among uninsured likely reflects underdiagnosis, not reduced risk.

Table 4

CKD Prevalence by Age Group

Age Group	2012	2021
18–44	7.8%	9.2%
45–64	14.9%	15.8%
65+	38.3%	36.6%

Note. Seniors show the highest burden, but Medicare may contribute to slight improvements.

Table 5

CKD Prevalence by Race × Age (65+)

Group (Age ≥65)	2012	2021
White	37.1%	35.4%
Black	41.6%	39.8%
Hispanic	39.2%	37.7%
Asian/Pacific Islander	33.9%	32.1%

Note. Seniors of color experience consistently higher CKD prevalence, revealing enduring inequities in chronic care access even within Medicare-eligible groups.

Table 6

https://doi.org/10.52202/081537-0001

CKD Prevalence by Income × Age (65+)

Income Group	2012	2021
Below Poverty Line (<1.0)	40.9%	39.7%
Above Poverty Line (≥1.0)	35.4%	33.6%

Note. Even among older adults, income remains a powerful determinant of CKD burden. The gap narrowed slightly, suggesting that infrastructure in higher-resource counties may support better outcomes.

Table 7

Uninsured Population by Age Group (2023 Estimates)

Age Group	Uninsured Rate (%)	Notes
Children (0–18)	~5.3%	Medicaid/CHIP coverage is prevalent
Adults (19–64)	~11.1%	Largest uninsured cohort
Seniors (65+)	~1.1%	Nearly all are covered by Medicare

Note. Adults 19–64 are most vulnerable. Underrepresentation in data reflects diagnostic barriers.

5. Discussion

5.1 Medicaid Expansion and Coverage Impact

The findings show that Medicaid expansion is associated with lower uninsured rates and greater detection of CKD, particularly among working-age adults. Expansion counties achieved larger reductions in uninsured rates for adults aged 19–64, and the increase in CKD prevalence among insured individuals suggests that expanded coverage improves diagnostic visibility *(KFF, 2024)*. From a leadership standpoint, expansion policies function as project-level levers: they increase access and facilitate early diagnosis. Public health leaders should advocate for continued expansion while pairing policy changes with investments in diagnostic capacity and outreach. Public health leaders operating in low-resource environments may benefit from adaptive budgeting models that reallocate staff time toward outreach and screening, particularly in high-burden communities (Bailey & Duru, 2022).

5.2 Intersectional Risk and Data Visibility

Cross-stratified analyses highlight the convergence of race, age and income in shaping CKD burden. Low-income seniors of color experience prevalence rates near 40 percent. These results imply that eligibility alone does not eliminate disparities. Infrastructure underinvestment and cultural disconnects in care delivery limit effectiveness. Leaders can use county-level dashboards and spatial analytics to identify clusters of vulnerability and direct resources to communities where demographic risks overlap. Adaptive systems thinking offers a useful framework for project planning.

5. 3 Infrastructure as Amplifier

Counties with Federally Qualified Health Centers, outreach programs and rural health networks reported sharper declines in CKD burden following Medicaid expansion. Infrastructure acts as an amplifier of policy success rather than a passive backdrop. Health systems and public health project leaders should integrate infrastructure development with coverage expansion. In non-expansion states, alternative funding and community partnerships can help build capacity until policy changes occur (Gaglioti et al., 2021).

5.4 Underdiagnosis and Data Interpretation

Lower reported CKD rates among uninsured adults reflect underdiagnosis more than genuine protection. Reduced contact with health services leaves uninsured individuals invisible in administrative datasets. Project leaders must interpret descriptive statistics with caution and ask whether low prevalence reflects true protection or underdiagnosis. When data suggest unexpectedly low disease rates among underserved groups, investments in screening and outreach should follow.

5.5 Limitations and Future Directions

This analysis is observational and does not establish causal relationships. County-level aggregation may mask within-county variation, and missing data for race and income in some rural regions limit full interpretation. These limitations, however, highlight opportunities. Public health systems can improve data infrastructure, promote standardized CKD screening and implement real-time analytics to monitor equity metrics. Future research might employ multilevel modeling to explore causal pathways and examine how specific infrastructure investments modulate the effects of Medicaid expansion.

6. Conclusion

Medicaid expansion serves as an equity lever by lowering uninsured rates and enhancing CKD detection across more than 2,500 U.S. counties. Yet non-Hispanic Black and Hispanic adults, older adults and low-income populations remain disproportionately affected. The intersection of race, age and income compounds vulnerability, even among Medicare-eligible populations. Infrastructure such as health centers and outreach programs amplifies the benefits of coverage expansion. Leaders should pair policy changes with investments in diagnostic capacity, culturally responsive programs and targeted outreach. Adaptive project management grounded in local data will be important for translating coverage gains into equitable health outcomes.

https://doi.org/10.52202/081537-0001

References

Bailey, S. R., & Duru, O. K. (2022). Medicaid expansion and access to nephrology care: Lessons from expansion and non-expansion states. *Journal of General Internal Medicine, 37*(9), 2265–2272.

Berkowitz, S. A., Wexler, D. J., & Baggett, T. P. (2019). *Systems strategies to address chronic disease disparities. Health Affairs, 38*(8), 1365–1372. https://doi.org/10.1377/hlthaff.2019.00390

Centers for Disease Control and Prevention. (2024). *Mapping poverty and chronic kidney disease among Medicare beneficiaries. Preventing Chronic Disease, 21*, Article 230286. https://www.cdc.gov/pcd/issues/2024/23_0286.htm

Centers for Disease Control and Prevention. (2025). *National Health Statistics Report No. 194: Insurance coverage and access to care by Medicaid expansion status.* https://www.cdc.gov/nchs/data/nhsr/nhsr194.pdf

Crews et al. (2023). Social vulnerability and chronic kidney disease outcomes across U.S. counties. *American Journal of Kidney Diseases, 81*(2), 210–219. https://doi.org/10.1053/j.ajkd.2022.08.016

Gaglioti, A. H., Petterson, S., Bazemore, A., & Phillips, R. L. Jr. (2021). *Health Equity Framework: Applying public health leadership to address systemic inequities. American Journal of Public Health, 111*(5), 806–813. https://doi.org/10.2105/AJPH.2021.306232

Kaiser Family Foundation. (2024). *The effects of Medicaid expansion under the ACA: Updated findings from a literature review*. https://www.kff.org/report-section/the-effects-of-medicaid-expansion-under-the-aca-updated-findings-from-a-literature-review-report/

U.S. Renal Data System. (2024). *Annual data report: Chronic kidney disease in the general population*. National Institute of Diabetes and Digestive and Kidney Diseases. https://usrds-adr.niddk.nih.gov/2024/chronic-kidney-disease/1-ckd-in-the-general-population

U.S. Census Bureau. (2024). *Small Area Health Insurance Estimates (SAHIE).* https://www.census.gov/programs-surveys/sahie.html

Wen, X., Liu, Z., Qiu, F., Leavitt, K., Wang, X., & Tang, Z. (2025). A power dependence model of the impact of leader impostorism on supervisor support and undermining: The moderating role of power distance. *Journal of Applied Psychology, 110*(7), 963–978. https://doi.org/10.1037/apl0001265

Investigating the Root Causes of Scope Creep in Oil and Gas Infrastructure Projects: A Systematic Literature Review

Felix Olufemi Ogele[a]

[a] *College of Professional Studies (CPS), Northeastern University, Toronto, Canada*

Corresponding author: ogele.f@northeastern.edu

Abstract

This research conducted a PRISMA-based systematic literature review (SLR) of the main drivers of scope creep in oil and gas infrastructure projects, which severely impacts project success. It covers scope creep's main causes, organizational and stakeholder dynamics, initial project definitions, and recommended practices to limit its consequences. Only 21 peer-reviewed journals published between 2019 and 2024 were used for full-text review and qualitative synthesis from 148 sources in Google Scholar, Emerald Insight, IEEE Xplore, ScienceDirect, and Scopus. The findings show that scope creep is caused by project requirements and specifications ambiguity, client revisions, insufficient scope definition, and low stakeholder participation. Clear scope declarations and engineering completion rating index reduce project uncertainty and scope revisions. The study recommends methodical planning and frequent communication to track scope changes and stay on track, like high-performing projects. Strategic management and recognizing scope creep's many causes can improve oil and gas infrastructure project outcomes, the study found. Organizational processes, specific contract conditions, and rigorous demand analysis reduce scope difficulties in complicated projects. The research is expected to assist project managers in enhancing performance to alleviate scope creep.

Keywords: Scope creep, project management, cost overruns, risk factors, stakeholder engagement, change management

https://doi.org/10.52202/081537-0002

I. Introduction

Oil and gas infrastructure projects refer to the various facilities and systems designed to support a wide array of activities in the entire lifecycle of oil and gas production and distribution. This includes exploration and drilling rigs, pipelines, refineries, gas processing plants, and storage tanks (Omara et al., 2023). These infrastructures are essential for transporting the resources from extraction sites to consumers, thus pivotal in ensuring energy security and economic stability in many countries. Despite their importance, oil and gas infrastructure projects face many challenges, including financial, environmental, and scope creep.

Scope creep, defined as the uncontrolled expansion of project scope without adjustments to resources, time, and budget, often leads to project overruns, and failure is a multifaceted issue in project management and is prevalent in oil and gas infrastructure projects. This phenomenon can be significantly related to the project performance domains outlined in the PMBOK 7th Edition, which emphasizes a performance-driven approach and recognizes the importance of various project domains that interact and affect project success (PMI, 2021). The eight project performance domains identified in the PMBOK 7th Edition are Stakeholders, Team, Development Approach and Lifecycle, Planning, Project Work, Delivery, Measurement, and Uncertainty and Ambiguity (Abdulla & Al-Hashimi, 2019; PMI, 2021). Each domain is critical in enhancing our understanding of the root causes and managing scope creep.

Most of the extant literature is concerned with the causes of scope creep in oil and gas infrastructure projects and seek to identify some of the dimensions that make scope creep so difficult to mitigate, including client change, poor scope definition, lack of stakeholder engagement, poor requirement analysis, and lack of an established change control system, among others. Cultural factors in organizational contexts also profoundly affect project scope management approaches, especially in the oil and gas industry, which frequently navigates diverse regulatory expectations across jurisdictions. Braam and Peeters (2017) assert that companies in environmentally sensitive sectors face heightened scrutiny about sustainable practices, which subsequently affects project expectations and scope. The interaction between local cultural norms and compliance standards requires project leaders to formulate culturally informed ways to address scope concerns successfully. However, none have been able to consider aligning the issue with the eight performance domains in the PMBOK 7th Edition. Understanding these causes and their connections to the various project domains is crucial to developing strategies to minimize their effects and foster successful project delivery.

This research aims to deepen and extend our understanding of the root causes of scope creep in oil and gas infrastructure projects and how the challenge can be mitigated by exploring these dynamics and synthesizing insights from relevant literature in addressing the following research questions: What are the principal root causes of scope creep in oil and gas infrastructure projects? What impact do organizational, contractual, and stakeholder-related issues have on scope creep? How significant is the initial project definition in mitigating scope creep? Finally, what best practices can be borrowed from high-performing projects to manage scope effectively?

II. Method

This study adopted a Systematic Literature Review (SLR) with an approach known as the Preferred Reporting Items for Systematic reviews and Meta-Analyses (PRISMA) method. The SLR seeks to answer specific research questions through a methodical selection and evaluation of the literature, aiming to minimize bias while synthesizing results (Saleh et al., 2023; Feng, 2023). The structured process within SLRs, which often adheres to established protocols like PRISMA, can help researchers address common issues of incomplete reporting that compromise the interpretability and reliability of systematic reviews (Page et al., 2021) and enables researchers to rigorously critique and summarize existing evidence, thereby identifying gaps in knowledge that can guide future research endeavors (Farooq et al., 2024). The SLR, as outlined by Page et al. (2021), was conducted in four stages: identification, screening, eligibility, and inclusion.

Identification Stage

This initial phase involves systematically searching for literature by utilizing predefined search strategies across various databases and sources. All potential studies are obtained, providing a large pool of literature for further examination. The following databases were chosen for the identification phase, on account of their relevance and academic rigour: Google Scholar, Emerald Insight, Scopus, IEEE Xplore, and Science Direct. Peer-reviewed journals written in English from all years until 2024 were considered. The search terms were modified together with informatics and combined with Boolean operators as follows, with the terms scope creep* or scope creep and oil infrastructure* or oil and gas infrastructure projects* or oil and gas infrastructure project scope* or oil and gas infrastructure challenge as keywords to identify as many eligible studies as possible. During the literature search, inclusion criteria (I1 to I4) were implemented, limiting the search to papers that contained specific keywords, represented original research, were designated as peer-reviewed, and were published in English (Table 1). Eventually, 148 peer-reviewed journal articles were retrieved.

Table 1: Criteria for inclusion and exclusion employed in the systematic review of peer-reviewed articles concerning for investigating the root causes of scope creep in oil and gas infrastructure projects.

Criteria	**Included (I)/Excluded (E)**
I1 - Papers containing keywords	Papers included
I2 - Peer-reviewed papers	Papers included
I3 - Original research	Papers included
I4 - Papers published in English language	Papers included
E1 - Duplicate papers	Papers excluded
E2 - Review papers, grey literature, thesis, and conference proceedings, grey literature	Papers excluded
E3 – Papers without full texts	Papers excluded
E4 - Studies that did not meet predefined inclusion criteria	Papers excluded
E5 - Papers published in a language other than English	Papers excluded

Screening Stage

In this stage, duplicates and studies that had not met predefined inclusion criteria were removed. The focus was on literature screening based on a review of the index and evaluating titles, abstracts, and keywords (TAK) to determine whether the studies merit further scrutiny based on their relevance to the research questions (Page et al., 2021; Cheng, 2023). In this process, literature that cannot be accessed in full text did not pass the screening stage, and the literature that successfully passed this stage was forwarded to the stage of the feasibility assessment process. After duplicates had been removed, title and abstract screening was conducted, and only 59 articles were taken to the next stage of the review process.

Eligibility Stage

The remaining 59 studies were further assessed for eligibility based on full-text reviews of the entire literature that had passed the screening stage against specific inclusion and exclusion criteria (Table 1). This critical stage

https://doi.org/10.52202/081537-0002

ensures that only studies that align closely with the research objectives are included in the final analysis (Page et al., 2021). Articles whose full texts were not available were excluded, and after full-text reviews, 38 articles were excluded.

Inclusion Stage

Finally, the remaining 21 journal articles whose year of publication period falls within 2019 and 2024 that met all predefined criteria were included in the systematic review, as shown in a flowchart (Figure 1) that outlines the study selection process, along with the final number of studies included for detailed analysis and synthesis. Figure 1 explains the identification process based on the search string and the inclusion criteria used, and then obtained 145 records with library details, namely Google Scholar, Emerald Insight, Scopus, IEEE Xplore, and Science Direct. Furthermore, 123 works of literature were obtained after checking for duplication.

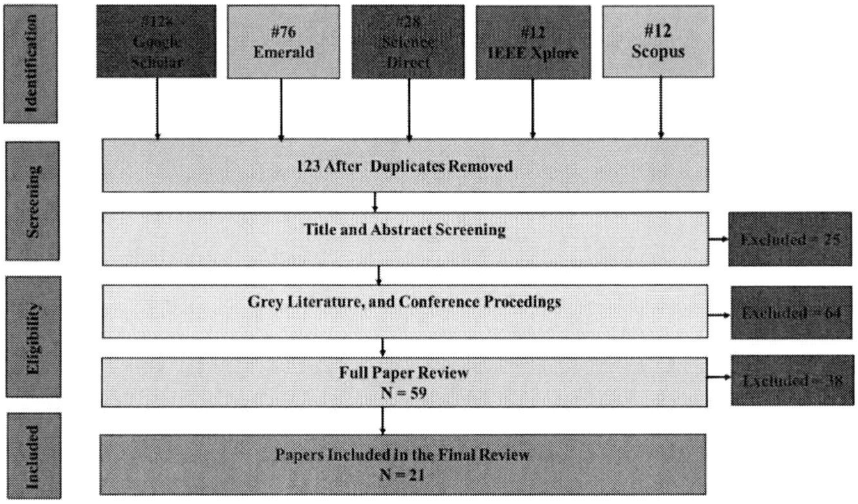

Figure 1: SLR Flowchart based on PRISMA

Table 2 shows the distribution of selected articles per year from 2019 to 2024. While 2019, 2020, 2021 and 2022 shared 19.05% each, 2024 shared 14.28% of the distribution. The remaining 9.52% was accounted for in 2023. This clearly shows that many papers discussed scope creep in oil and gas infrastructure projects between 2019 to 2022.

Table 2: Distribution of Selected Study Articles Per Year

Year	Frequency	Percentage
2019	4	19.05
2020	4	19.05
2021	4	19.05
2022	4	19.05
2023	2	9.52
2024	3	14.28
Total	21	100

Details of the 21 peer-reviewed papers reviewed in this study are shown in Table 3, consisting of the year, source, title, and citation.

Table 3: Details of Papers Reviewed

S/N	Year	Source	Title	Author(s)
1.	2019	Scopus	The impact of project management methodologies on project success: a case study of the oil and gas industry.	Abdulla & Al-Hashimi (2019).
2.	2019	Emerald Insight	Exploring factors behind project scope creep – stakeholders' perspective.	Ajmal et al. (2019)
3.	2019	Google Scholar	Time and Cost Overrun in the Saudi Arabian Oil and Gas Construction Industry.	Seddeeq et al., (2019)
4.	2019	Google Scholar	Earned value analysis, implementation barriers, and maturity level in oil & gas production.	Zhan et al. (2019)
5.	2020	Emerald Insight	Theoretical review on critical risk factors in oil and gas construction projects in Yemen.	Kassem et al. (2020a)
6.	2020	Emerald Insight	Structural modelling of internal risk factors for oil and gas construction projects.	Kassem et al. (2020b)
7.	2020	Emerald Insight	Multi-objective mathematical model based on fuzzy hybrid multi-criteria decision-making and FMEA approach for the risks of oil and gas projects.	Khalilzadeh et al. (2020)
8.	2020	IEEE Xplore	The impact of scope creep on project success: an empirical investigation.	Komal et al. (2020)
9.	2021	Emerald Insight	Managing project scope creep in construction industry.	Ajmal et al. (2021)
10.	2021	Google Scholar	Incidents and Disaster Avoidance: The Role of Communication Management and the Organizational Communication Climate in High-Risk Environments	Nordin et al. (2021)
11.	2021	IEEE Xplore	An Empirical Investigation of Factors Causing Scope Creep in Agile Global Software Development Context: A Conceptual Model for Project Managers.	Aizaz et al. (2021)
12.	2021	Google Scholar	Predefined project scope changes and its causes for project success.	Althiyabi & Quereshi (2021)
13.	2022	ScienceDirect	Avoiding or disregarding: Exploring the relationship between scope creep, project complexity, and the success of construction projects.	Ahmed & Jawad (2022)
14.	2022	Google Scholar	Critical factors to overcome project scope creep in the oil and gas industry.	Alkarbi et al. (2022)
15.	2022	Emerald Insight	A review on prospective risks and mitigation for oil and gas projects: implication for Indian CGD companies.	Rawat et al., (2022)
16.	2022	Google Scholar	Cost Overruns in Engineering Procurement Construction (EPC) Fabrication Oil and Gas Megaprojects in Malaysia: The Importance of Resource Allocation (5M)	Roshidi et al. (2022)
17.	2023	Google Scholar	Influence of Organizational Factors on Safety Performance in Oil and Gas Industry in the Niger Delta Region, Nigeria.	Okezie et al. (2023)

https://doi.org/10.52202/081537-0002

18.	2023	Emerald Insight	Stakeholder management challenges and strategies for sustainability issues in megaprojects: case studies from Australia.	Senaratne et al. (2023)
19.	2024	Google Scholar	Assessing the Influence of Various Work Breakdown Structures on Project Completion Time.	Odedairo (2024)
20.	2024	Google Scholar	Quantitative insights into risk management and project success: evidence from Malaysia's oil and gas sector.	Othman & Khan (2024)
21.	2024	Google Scholar	A functional vs projectized approach in oil and gas industry.	Khatib et al. (2024)

III. Results and Discussion

After executing the analysis as outlined in the Methods section, the literature dataset of 15 articles were clustered into 10 themes including (i) Inadequate Project Scope Definition; (ii) poor requirements gathering; (iii) weak change management process; (iv) stakeholder influence issues; (v) ineffective communication; (vi) Governance leadership failure; (vii) inadequate risk management; (viii) resource constraints estimation; (ix) contract and procurement issues; and (x) external environmental factors. These 10 themes, as shown in Table 4, provide insight into the sample and reveal the underlying causes of scope creep in oil and gas infrastructure projects. Aligning the themes with the eight performance domains further deepens our understanding of the various areas and stages in the project lifecycle that must be considered to mitigate scope creep in oil and gas infrastructure projects, and by extension, in other construction projects as well.

Table 4

Scope Creep Theme	Issues	Citation	Aligned PM Domains	RQs Addressed
Inadequate Project Scope Definition	Poor scope definition.	Ahmed & Jawad, (2022); Seddeeq et al., (2019); Alkarbi et al., (2022); Kassem et al., (2020a); Fageha and Aibinu, (2014); Amoatey & Anson, (2017); Agho et al., (2023); & Khalilzadeh et al., (2020).	Planning, Delivery, Uncertainty	RQ1, RQ2, RQ3, & RQ4
	Inadequate understanding of scope.	Aizaz et al., (2021).		
	Lack of stakeholder engagement	Ahmed & Jawad, (2022); Ajmal et al., (2019); Ajmal et al., (2021); Alkarbi et al., (2022); Senaratne et al., (2023); Seddeeq et al., (2019); Anantatmula & Yang, (2013); Okezie et al., (2023); Othman & Khan, (2024); Khalilzadeh et al. (2020).		
Poor requirements Gathering	Poor requirement analysis.	Ahmed & Jawad, (2022); Roshdi et al., (2022).	Stakeholder, Planning, Development Approach	RQ1, RQ2, RQ3, & RQ4
	Poor initial requirement specifications	Ahmed & Jawad, 2022; Ajmal et al., 2021; Senaratne et al., 2023; Okezie et al., 2023; Othman & Khan, 2024; Khalilzadeh et al. (2020)		

Weak Change Management Process	Client change. Lack of an established change control system Organizational lack of project management frameworks. Absence of defined roles. Organizational insufficient focus on scope management	Amoatey & Anson, (2017); Ajmal et al., (2019); Seddeeq et al., (2019); Aizaz et al., (2021); Kassem et al., (2019); Alkarbi et al., (2022); Kazemi et al., (2018); Khalilzadeh et al. (2020); Okezie et al., (2023). Jayalath & Somarathna, 2021. Seddeeq et al., (2019); Komal et al., (2020); Abdulla & Al-Hashimi, (2019); Althiyabi & Qureshi, (2021); Odedairo, (2024). Barghi & Sikari, (2020). Maduri et al., (2018)	Project Work, Planning, Measurement	RQ1, RQ2, RQ3 & RQ4
Stakeholder Influence Issues	Design changes	Seddeeq et al., (2019); Alkarbi et al., (2022); Kazemi et al., (2018); & Khalilzadeh et al. (2020)	Stakeholder, Team, Uncertainty	RQ1, RQ2 & RQ4
Ineffective Communication	Poor communication	Ahmed & Jawad, (2022); Ajmal et al., (2021); Alkarbi et al., (2022); Okezie et al., (2023); Khalilzadeh et al. (2020); Nordin et al. (2021); Agho et al., (2023); & Senaratne et al., (2023).	Stakeholder, Team, Project Work	RQ1, RQ4
Governance and Leadership Failures	Poor Planning. Managerial inadequacies. Mismanagement of stakeholder expectations. Unmet needs of stakeholders. Lack of transparency. Inefficient organizational structures	Ahmed & Jawad, (2022); Ajmal et al., (2019); Seddeeq et al., 2019; Kassem et al., (2019); Kassem et al., (2020a); Khalilzadeh et al. (2020). Jayalath & Somarathna, (2021); Khalilzadeh et al. (2020). Abdulla & Al-Hashimi, (2019); Komal et al., (2020); Nordin et al. (2021). Agho et al., (2023); Jayalath & Somarathna, (2021). Alkarbi et al., (2022); Okezie et al., (2023); & Nordin et al. (2021). Alkarbi et al., (2022); Okezie et al., (2023); Khalilzadeh et al. (2020)	Project Work, Team, Stakeholder	RQ1, RQ2, & RQ4
Inadequate Risk Management	Poor risk management practices. Poor performance measurement practices	Kassem et al., (2019); Khalilzadeh et al., (2020); Kassem et al., (2020a); Kassem et al., (2020b); Othman & Khan, 2024. Ahmed & Jawad, (2022); Zhan et al., (2019); Seddeeq et al., (2019); Kassem et al., (2020).	Uncertainty, Planning, Measurement	RQ2, & RQ4

https://doi.org/10.52202/081537-0002

Resource Constraints & Estimation	Corrupt practices. Inadequate forecasting. Constrained budgets	Alkarbi et al., (2022); Othman and Khan, (2024). Rawat et al., (2022). Kassem et al., (2019).	Planning, Project Work, Measurement	RQ1, RQ2, & RQ4
Contract and Procurement Issues	Contractual vagueness. Ineffective procurement strategies	Ajmal et al., (2019); Ajmal et al., (2021); Seddeeq et al., (2019); Alkarbi et al., (2022); & Senaratne et al., (2023); Roshdi, et al., (2022). Alkarbi et al., (2022); & Kassem et al., (2020b)	Delivery, Stakeholder, Planning	RQ1, RQ2, RQ3, RQ4
External Environmental Factors	Changes in regulations. Fluctuations in the economy and market dynamics. Ecological impacts and community concerns. Social pressures or community opposition.	Rawat et al., (2022); Seddeeq et al., (2019). Rawat et al., (2022); Othman & Khan, (2024). Kassem et al., (2020c). Othman & Khan, (2024).	Uncertainty, Planning, Development Approach	RQ1, RQ2, RQ4

Inadequate Project Scope Definition: A significant body of literature on scope creep is related to scope definition. The key findings under this theme that have been documented to be significant contributors to scope creep across multiple sources are inadequate understanding of scope, poor scope definition, and lack of stakeholders' engagement or involvement in defining the scope of a project. These findings align with the existing research, which recognizes scope definition as a pivotal success factor in project management (PMI, 2021). The study validates that comprehensive scope planning—integrating stakeholder feedback, feasibility assessment, and precise documentation—is crucial for project stability and success.

Poor requirements gathering: This theme is pivotal as a catalyst for scope creep and a driver of scope creep in oil and gas infrastructure projects. The discovery indicates that the failure to elicit, document, and validate stakeholder requirements successfully in the initial phases of a project results in misaligned expectations and repeated alterations during project execution.

Weak Change Management Process: This theme highlights the lack of defined protocols for managing project modifications, despite their inevitability arising from shifting site conditions, regulatory alterations, and technological advancements. This study finds that an inadequate change management process catalyzes scope creep, as evidenced by client alterations, the absence of a formal change control system, deficiencies in organizational project management frameworks, a lack of clearly defined responsibilities, and an inadequate emphasis on scope management within the organization.

Stakeholder influence issues: Stakeholder influence is a significant factor in scope creep throughout oil and gas infrastructure projects. The reviews show that the stakeholder influence issue occurs due to various reasons, including inadequate stakeholder communication, client-driven changes, ambiguity in roles and responsibilities, and the influence of multiple stakeholders.

Ineffective Communication: Ineffective communication as a theme is a critical issue that is a contributor to scope creep in oil and gas infrastructure projects. This shows that a lack of clear, timely, and transparent communication among stakeholders often leads to misunderstandings, misaligned expectations, and ultimately, project scope changes.

Governance Leadership Failure: Failures in governance or organizational leadership contribute to scope creep and can markedly intensify scope creep in oil and gas infrastructure projects. The finding shows how the amalgamation of deficient leadership, absence of explicit guidelines, poor organizational culture, insufficient stakeholder involvement, and neglect in addressing stakeholder impact fosters an atmosphere that facilitates the extension of project scope beyond its original limitations.

Inadequate Risk Management: Inadequate risk management procedures substantially exacerbate scope creep. This study indicates that deficiencies in risk management stem from multiple causes, including inadequate planning, poor stakeholder participation, inadequate feasibility studies, poor project cost estimation, and an inability to foresee unforeseen dangers.

Resource Constraints Estimation: Resource constraints estimation is essential in project management for oil and gas infrastructure projects, as it directly affects planning, execution, and ultimate success. Inadequate estimation or management of resource limits might result in substantial risk factors that contribute to scope creep. This study finds that resource constraints estimation can manifest in various ways, including inadequate forecasting, constrained budget, and corrupt practices.

Contract and Procurement Issues: The interplay between contract and procurement challenges and scope creep in oil and gas infrastructure projects is complex and nuanced. As seen in this study, these concerns can cause scope creep and profoundly affect project timetables, finances, and overall success.

External Environmental Factors: This 10th theme captures the literature that addresses risk and uncertainty that transcends the scope and organizations involved in a project that contribute to scope creep. This, as discovered in the study, could be a result of many factors, including fluctuations in economic conditions and market forces, changes in regulations, ecological factors and community concerns, security issues, and geopolitical instability.

IV. Discussion

This study has synthesized 10 core themes that illuminate the root causes of scope creep in oil and gas infrastructure projects. The findings confirm that scope creep is not only a technological error but a complex management issue rooted in organizational structures, stakeholder interactions, and external uncertainty. This discussion not only reiterates the descriptive results but also examines their strategic implications and offers practical recommendations to enhance project execution in complex infrastructure settings.

Scope and Requirements Definition: Strategic Foundation for Success

The prevalence of insufficient scope definition and inadequate requirements gathering highlights the necessity for thorough front-end planning. This assertion is corroborated by Satria et al. (2023), who demonstrated how effective techniques in the front-end planning phase, including precise project scope definition and thorough needs evaluations, significantly diminish the probability of project delays and budget overruns. Project teams should regard scoping not as a singular documentation task, but as a collaborative and iterative endeavor. Incorporating requirements engineering approaches, performing preliminary feasibility evaluations, and employing organized stakeholder engagement frameworks, such as workshops and design charrettes, can

https://doi.org/10.52202/081537-0002

markedly diminish the likelihood of misalignment and subsequent scope alterations. Institutionalizing these principles improves project predictability and strengthens responsibility.

Change Management: From Reactive to Proactive

The persistent deficiency in change management underscores the systemic lack of codified protocols. Oil and gas sector projects must transcend ad hoc reactions by including a comprehensive change control system in accordance with industry best practices, such as those delineated in PMBOK or AACE recommendations. Proactive change management must incorporate impact assessment processes, define approval roles, and establish real-time communication channels, enabling project teams to assimilate change without undermining fundamental project objectives. The project plans may encompass the implementation of agile approaches, which can substantially assist project teams (Gräßler et al., 2022). This change management strategy corroborates the previous conclusions of Cullen & Parker (2015), which argued that a proactive methodology can potentially augment project resilience to scope alterations and improve overall performance.

Stakeholder Dynamics: Navigating Complexity with Clarity

The impact of diverse stakeholders with differing priorities underscores the necessity of early and ongoing stakeholder mapping and engagement. Instead of viewing stakeholder involvement as a potential source of scope risk, project leaders ought to leverage it as a governance instrument. Utilizing stakeholder influence matrices, fostering clear communication, and establishing specified decision rights can aid in reconciling opposing interests and reducing unauthorized scope alterations. Involving a variety of pertinent stakeholders from the first stages is supported by Aapaoja & Haapasalo (2014), who regard this as essential for mitigating potential uncertainties and difficulties that could result in scope deviations later in the project lifecycle. This further corroborates the findings of Baharuddin et al. (2021), which indicated that early engagement might cultivate a culture of partnership and collaboration, alleviating risks related to unmet stakeholder expectations and thereby diminishing the probability of scope creep.

Governance and Leadership: The Enabler of Scope Discipline

Research on governance failures indicates that corporate culture and leadership alignment are crucial for effective scope management. Leadership must advocate for a culture of discipline, guarantee consistency in communication, and implement standardized scope and change management methods. Robust governance frameworks, rooted in clearly articulated project charters, matrices of delegated authority, and performance dashboards, can improve supervision and mitigate leadership-induced deviations in project scope. Previous research corroborates the premise that strong governance and leadership are essential facilitators in managing scope discipline, especially in oil and gas infrastructure projects. Kim & Kim (2012) contend that leadership not only governs processes but also influences social dynamics within project teams, hence fostering a culture of accountability and clarity regarding project scopes.

Communication and Risk Integration: The Cornerstones of Resilience

Communication failures and inadequate risk integration frequently overlap, exacerbating the consequences of scope creep. Future initiatives must incorporate communication planning within risk management, acknowledging that delayed or erroneous messages can transform modest risks into significant disruptions. By associating communication protocols with risk registers, projects can guarantee that emergent dangers are recognized, deliberated, and managed prior to requiring reactive scope alterations. This supports the recommendation of Khalilzadeh et al. (2020) that implementing multi-criteria decision-making approaches can assist businesses in effectively managing diverse risk dimensions, so ensuring that projects stay on course despite the obstacles posed by scope creep.

Procurement and Contractual Agility

Contractual and procurement challenges indicate the necessity for adaptive contracting techniques. Conventional fixed-scope contracts may be inappropriate for contexts characterized by regulatory instability and shifting stakeholders. Hybrid contracts, which integrate fixed-price components with incentive or alliance-based frameworks, may offer the necessary flexibility to address changing project requirements while maintaining scope integrity. The significance of adaptive contracting is in its capacity to enable flexibility in addressing changing project requirements. This corresponds with the findings of Hurmerinta-Haanpää & Viding (2018), which emphasize that contracts fulfill diverse roles in interorganizational relationships, allowing parties to modify their practices in response to evolving conditions.

Resource Planning and External Environmental Preparedness

Underestimating resource limitations and external environmental threats necessitates improved forecasting ability. This encompasses technical estimating methods, scenario planning, and risk simulations to evaluate project resilience under varying resource and geopolitical scenarios. Incorporating foresight during the initial planning phases can significantly mitigate scope creep caused by reactive reallocations or crisis-induced modifications. Strengthening the resilience of oil and gas infrastructure against unforeseen external problems not only facilitates efficient resource planning but also promotes sustained adherence to legislation aimed at safeguarding environmental interests (Mahmood et al. 2023).

Implications for Practice and Policy

The study's findings yield various practical implications essential for enhancing project performance and reducing scope creep in oil and gas infrastructure projects. These implications guide both organizational practices and policy actions aimed at enhancing project governance, planning precision, and stakeholder alignment. The subsequent guidelines delineate essential focal points for practitioners, sponsors, and regulatory authorities:

Project sponsors and managers in the oil and gas sector must allocate resources for capacity building in scope and change management, ensuring coherence across project planning and execution teams. Regulatory organizations and funding agencies must require comprehensive scope and stakeholder management plans as conditions for project approval. Organizations should establish and implement standard operating procedures for managing scope creep, based on industry principles yet adapted to sector-specific conditions.

Strategies to Address Scope Creep in Oil and Gas Infrastructure Projects

Clear and Comprehensive Project Definition: Limiting scope creep requires a clear project scope. Early stakeholder participation during project conception is essential to capture all project requirements and expectations. Amoatey and Anson (2017) state that early stakeholder involvement improves project scope definition, preventing miscommunication and execution changes. This proactive participation can greatly improve project boundaries and deliverables knowledge. A precise project scope statement produced with stakeholders early on is crucial. To ensure stakeholder awareness, this planning process should include explicit project needs and restrictions (Rawat et al., 2022).

Establishing Rigorous Change Control Procedures: A strong change control process is necessary to manage scope changes. Every modification should be evaluated for its impact on project timeframes and costs. Khalilzadeh et al. (2020) recommend using multi-criteria decision-making (MCDM) and FMEA to assess change risks. Project managers can tighten scope control by evaluating each modification request before approval. A systematic change control procedure lets project managers assess proposed modifications' effects on timelines and budgets before accepting them. Khalilzadeh et al. (2020) recommend qualitative and quantitative studies to determine change feasibility. Implementing strict change control methods can assist in evaluating and managing proposed changes.

https://doi.org/10.52202/081537-0002

This technique lets project teams assess the effects of changes on deadlines and budgets before making decisions (Komal et al., 2020; Zhan, 2019).

Clear Documentation and Scope Management: Maintaining comprehensive documentation throughout the project lifecycle aids in avoiding scope creep. Documenting scope statements, requirements, and change requests ensures stakeholder alignment with project goals. Shirazi et al. (2017) stress the importance of precise project scope documentation to help manage scope modifications by providing a reference point for what was originally agreed upon. This strategy reduces project expectations miscommunications. A well-structured scope management plan should govern scope modifications throughout the project. This plan should define deliverables and stakeholder roles to avoid confusion (Ajmal et al., 2019).

Training and Empowering Project Teams: Training and developing project management skills can improve team performance and decrease scope creep. Irfan et al. (2019) found that firms that invest in project management maturity have superior project outcomes, including scope creep management. Training personnel in setting limits, engaging with stakeholders, and following the change management process helps them tackle scope change challenges. Training project managers and teams to recognise scope creep risks and use best practices in project delivery can reduce their impact (Ajmal et al., 2019; Abdulla & Al-Hashimi, 2019). Training project teams in scope and risk management improves their scope creep management. Project management skills help teams handle oil and gas project complexity (Othman & Khan, 2024).

Implementation of Earned Value Management (EVM): Using Earned Value Management as a systematic way to track project progress against scope, time, and cost. EVM helps project managers identify scope deviations early and take corrective action (Zhan et al., 2019). EVM quantifies project performance to manage stakeholder expectations and avoid scope creep.

Ongoing Communication and Engagement: Continuous communication with stakeholders throughout the project is essential for alignment and expectation management. Regular updates and discussions can address concerns before they cause major changes (Seddeeq et al., 2019; Othman & Khan, 2024). Nordin et al. (2021) assert that high-performing projects should prioritize proactive communication planning, foster an open and trust-based communication environment, and implement cross-functional coordination and feedback methods. These methods augment responsiveness, boost stakeholder alignment, and facilitate the early identification and settlement of potential scope deviations, hence constituting best practices for scope control in intricate oil and gas settings.

Role of Project Management Offices: Creating a PMO improves project planning and execution. A PMO centralizes project knowledge and tools to ensure alignment with organizational goals and best practices. Structured project management practices through PMO initiatives can reduce scope-related risks and improve project success, according to Aizaz et al. (2021). PMOs can improve project communication and scope management across the company. Agho et al. (2023) identify real-time monitoring, predictive analytics, and scenario modelling as essential advantages of digital twins, as evidenced in high-performing energy projects. PMOs can use these techniques as exemplary strategies to foresee and manage scope alterations prior to their escalation, facilitate proactive stakeholder involvement via data-sharing dashboards, and enhance dynamic contract and risk management grounded in real-time performance metrics.

V. Conclusion

This study reconceptualizes scope creep not just as a symptom of inadequate planning but as a consequence of deficiencies in governance, communication, and systemic structures, hence prompting a reconsideration of project management strategies within the oil and gas sector. Future interventions should transition from corrective measures to adopt preventative tactics, founded on transparency, stakeholder alignment, and procedural rigor.

What is missing in the current research is an understanding of how factors like local regulations and cultural attitudes toward change management can influence scope stability. This shows some areas remain largely unexplored or require further probing. There is a need for comparative studies that will assess scope creep incidences in oil and gas infrastructure projects across different countries in the global south and the global north to better understand how factors like local regulations and cultural attitudes toward change management can influence scope stability from planning to delivery. Future research should examine how scope creep affects different types of construction projects across various geographical locations and sectors (Ajmal et al., 2021).

Most current literature employs cross-sectional data analysis, leaving a gap in understanding the scope creep's effect over project lifecycles, which helps to inform better preventative measures. There is therefore a need for longitudinal studies that will reveal how scope creep impacts project outcomes over time, identify patterns and factors contributing to scope creep in the oil and gas sector, and the effectiveness of various mitigation strategies. Researchers might adopt a longitudinal approach to analyze the long-term effects of implemented scope management strategies on project success (Ajmal et al., 2021).

More research needs to be undertaken on the roles of stakeholders and how their often-conflicting interests, needs, and priorities contribute to scope creep in oil and gas infrastructure projects.

https://doi.org/10.52202/081537-0002

References

Aapaoja, A. and Haapasalo, H. (2014). A framework for stakeholder identification and classification in construction projects. *Open Journal of Business and Management, 02*(01), 43-55. https://doi.org/10.4236/ojbm.2014.21007 Google Scholar

Abaza, B., & Kisi, K. (2024). A lesson learned from construction projects to prevent scope creep. *Proceedings of 60th Annual Associated Schools of Construction International Conference, 5*, 659-667. https://doi.org/10.29007/zmdp Google Scholar

Abdulla, H., & Al-Hashimi, M. (2019). The impact of project management methodologies on project success: a case study of the oil and gas industry. *Journal of Engineering Project and Production Management.* https://doi.org/10.2478/jeppm-2019-0013 SCOPUS

Adam, A., Josephson, P., & Lindahl, G. (2017). Aggregation of factors causing cost overruns and time delays in large public construction projects. *Engineering Construction & Architectural Management, 24*(3), 393-406. https://doi.org/10.1108/ecam-09-2015-0135 Emerald Insight

Ahmed, R., & Jawad. M. (2022). Avoiding or disregarding: Exploring the relationship between scope creep, project complexity, and the success of construction projects. *Project Leadership and Society, 3*,100064 https://doi.org/10.1016/j.plas.2022.100064 ScienceDirect

Aizaz, F., Khan, S., Khan, J., Rehman, I., & Akhunzada, A. (2021). An Empirical Investigation of Factors Causing Scope Creep in Agile Global Software Development Context: A Conceptual Model for Project Managers. *IEEE Access, 9*, 109166-109195. https://doi.org/10.1109/access.2021.3100779 IEEE Xplore

Ajmal, M., Khan, M., & Al-Yafei, H. (2019). Exploring factors behind project scope creep – stakeholders' perspective. *International Journal of Managing Projects in Business, 13*(3), 483-504. https://doi.org/10.1108/ijmpb-10-2018-0228 Emerald Insight

Ajmal, M., Khan, M., Gunasekaran, A., & Helo, P. (2021). Managing project scope creep in construction industry. *Engineering Construction & Architectural Management, 29*(7), 2786-2809. https://doi.org/10.1108/ecam-07-2020-0568 Emerald Insight

Alkarbi, A., Ajmal, M., & Salameh, A. (2022). Critical factors to overcome project scope creep in the oil and gas industry. *International Journal of Service Science Management Engineering and Technology, 13*(1), 1-21. https://doi.org/10.4018/ijssmet.298674 Google Scholar

Althiyabi, T. & Qureshi, M. (2021). Predefined project scope changes and its causes for project success. *International Journal of Software Engineering & Applications, 12*(3), 45-56. https://doi.org/10.5121/ijsea.2021.12304 Google Scholar

Amoatey, C. & Anson, B. (2017). Investigating the major causes of scope creep in real estate construction projects in Ghana. *Journal of Facilities Management, 15*(4), 393-408. https://doi.org/10.1108/jfm-11-2016-0052 Emerald Insight

Anantatmula, V. & Yang, F. (2013). Risk management instruments, strategies and their impact on project success. International Journal of Risk and Contingency Management, 2(2), 27-41. https://doi.org/10.4018/ijrcm.2013040102 Google Scholar

Baharuddin, H. E. A., Azizi, N. S. M., & Piri, I. (2021). Improving early stakeholder engagement process: an empirical study of rapid transit railway of Malaysian project. *International Journal of Sustainable Construction Engineering and Technology, 12*(5). https://doi.org/10.30880/ijscet.2021.12.05.028 Google Scholar

Braam, G. and Peeters, R. (2017). Corporate Sustainability Performance and Assurance on Sustainability Reports: Diffusion of Accounting Practices in the Realm of Sustainable Development. *Corporate Social Responsibility and Environmental Management, 25*(2), 164-181. https://doi.org/10.1002/csr.1447 Google Scholar

Cullen, K. and Parker, D. (2015). Improving performance in project-based management: synthesizing strategic theories. *International Journal of Productivity and Performance Management, 64*(5), 608-624. https://doi.org/10.1108/ijppm-02-2014-0031 Emerald Insight

Emeka-Okoli, S., Nwankwo, T., Otonnah, C., & Nwankwo, E. (2024). Environmental stewardship and corporate social responsibility: a review of case studies from the oil and gas sector. *World Journal of Advanced Research and Reviews, 21*(3), 069-077. https://doi.org/10.30574/wjarr.2024.21.3.0661 Google Scholar

Feng, C. (2023). Studies on global remote interpreting: A PRISMA systematic review. In *SHS web of conferences,162*, 01031. EDP Sciences. https://doi.org/10.1051/shsconf/202316201031 Google Scholar

Fageha, M. and Aibinu, A. (2014). Prioritising project scope definition elements in public building projects. *Construction Economics and Building, 14*(3), 18-33. https://doi.org/10.5130/ajceb.v14i3.4155 Google Scholar

Farooq, F., Mohammad, S., Nazir, N., & Shah, P. (2024). Happiness at work: a systematic literature review. *International Journal of Organizational Analysis, 32*(10), 2236-2255. https://doi.org/10.1108/ijoa-08-2023-3902 Emerald Insight

Gräßler, I., Oleff, C., & Preuß, D. (2022). Proactive management of requirement changes in the development of complex technical systems. *Applied Sciences, 12*(4), 1874. https://doi.org/10.3390/app12041874 Google Scholar

Guo, X., An, G., Han, J., & Wang, Z. (2024). How can fund leaders utilize organizational networks to enhance organizational sustainable performance using the funds' co-holding network as a tool. *Sage Open, 14*(2). https://doi.org/10.1177/21582440241247644 Google Scholar

Hurmerinta-Haanpää, A. and Viding, S. (2018). The functions of contracts in interorganizational relationships: a contract experts' perspective. *Journal of Strategic Contracting and Negotiation, 4*(1-2), 98-118. https://doi.org/10.1177/2055563619884791 Google Scholar

Irfan, M., Hassan, M., & Hassan, N. (2019). The effect of project management capabilities on project success in Pakistan: An empirical Investigation. *IEEE Access, 7*, 39417-39431. https://doi.org/10.1109/access.2019.2906851 IEEE Xplore

Jayalath, C., & Somarathna, K. K. G. P. (2021). Key performance indicators in upholding scope creep management in road projects. In *Proceedings The 9th World Construction Symposium| July* (p. 382). Google Scholar

Kassem, M. A., Khoiry, M. A., & Hamzah, N. (2019). Risk factors in oil and gas construction projects in developing countries: a case study. *International Journal of Energy Sector Management, 13*(4), 846-861. https://doi.org/10.1108/ijesm-11-2018-0002 Emerald Insight

Kassem, M., Khoiry, M., & Hamzah, N. (2020a). Theoretical review on critical risk factors in oil and gas construction projects in Yemen. *Engineering Construction & Architectural Management, 28*(4), 934-968. https://doi.org/10.1108/ecam-03-2019-0123 Emerald Insight

Kassem, M., Khoiry, M., & Hamzah, N. (2020b). Structural modelling of internal risk factors for oil and gas construction projects. *International Journal of Energy Sector Management, 14*(5), 975-1000. https://doi.org/10.1108/ijesm-11-2019-0022 Emerald Insight

Kassem, M., Khoiry, M., & Hamzah, N. (2020c). Assessment of the effect of external risk factors on the success of an oil and gas construction project. *Engineering Construction & Architectural Management, 27*(9), 2767-2793. https://doi.org/10.1108/ecam-10-2019-0573 Emerald Insight

Khalilzadeh, M., Balafshan, R., & Hafezalkotob, A. (2020). Multi-objective mathematical model based on fuzzy hybrid multi-criteria decision-making and FMEA approach for the risks of oil and gas projects. *Journal of Engineering Design and Technology, 18*(6), 1997-2016. https://doi.org/10.1108/jedt-01-2020-0020 Emerald Insight

Khatib, M., Hammadi, A., Oraby, K., & Hamar, A. (2024). A functional vs projectized approach in oil and gas industry. *International Journal of Theory of Organization and Practice (IJTOP), 3*(2), 197-209. https://doi.org/10.54489/ijtop.v3i2.315 Google Scholar

Kim, J. and Kim, G. (2012). Analyzing the Priority of Leadership Elements for Project Manager of Building Construction Project. *Journal of the Korea Institute of Building Construction, 12*(1), 31-41. https://doi.org/10.5345/jkibc.2012.12.1.031

Komal, B., Janjua, U., Anwar, F., Madni, T., Cheema, M., Malik, M., & Shahid, A. (2020). The impact of scope creep on project success: an empirical investigation. *IEEE Access, 8*, 125755-125775. https://doi.org/10.1109/access.2020.3007098 IEEE Xplore

Madhuri, K., Mokashi, U., & Suma, V. (2018). A triangular perception of scope creep influencing the project success. *International Journal of Business Information Systems, 27*(1), 69-85. https://doi.org/10.1504/ijbis.2018.10009189 Google Scholar

Mahmood, Y., Afrin, T., Huang, Y., & Yodo, N. (2023). Sustainable development for oil and gas infrastructure from risk, reliability, and resilience perspectives. *Sustainability, 15*(6), 4953. https://doi.org/10.3390/su15064953 Google Scholar

Nordin, S., Rizal, A., Rashid, R., Omar, R., & Priyadi, U. (2021). Incidents and Disaster Avoidance: The Role of Communication Management and the Organizational Communication Climate in High-Risk Environments. *Sustainability, 13*(18), 10138. https://doi.org/10.3390/su131810138 Google Scholar

Odedairo, B. O. (2024). Assessing the Influence of Various Work Breakdown Structures on Project Completion Time. *Engineering, Technology & Applied Science Research,14*(2), 13773-13779. https://doi.org/10.48084/etasr.7023 Google Scholar

Okezie, E. O., Nwaogazie, I. L., & Chinemerem, P. (2023). Influence of Organizational Factors on Safety Performance in Oil and Gas Industry in the Niger Delta Region, Nigeria. *Archives of Current Research International, 23*(1), 15-23. https://doi.org/10.9734/acri/2023/v23i1551 Google Scholar

Omara, M., Gautam, R., O'Brien, M., Himmelberger, A., Franco, A., Meisenhelder, K., Hauser, G., Lyon, D. R., Chulakadabba, A., Miller, C. C., Franklin, J., Wofsy, S. C., & Hamburg, S. (2023). Developing a spatially explicit global oil and gas infrastructure database for characterizing methane emission sources at high resolution. *Earth Syst. Sci. Data, 15*, 3761-3790. https://doi.org/10.5194/essd-15-3761-2023 Google Scholar

Othman, N. and Khan, M. (2024). Quantitative insights into risk management and project success: evidence from Malaysia's oil and gas sector. *Malaysian Journal of Business Economics and Management, 3*(2), 144-149. https://doi.org/10.56532/mjbem.v3i2.69 Google Scholar

Panyard, D., Ramly, E., Dean, S., & Bartels, C. (2018). Bridging clinical researcher perceptions and health it realities: a case study of stakeholder creep. *International Journal of Medical Informatics, 110*, 19-24. https://doi.org/10.1016/j.ijmedinf.2017.11.014 Science Direct

Page, M., McKenzie, J., Bossuyt, P., Boutron, I., Hoffmann, T., Mulrow, C., Shamseer, L., Tetzlaff, J. M., Akl, E. A., Brennan, S. E., Chou, R., Glanville, J., Grimshaw, J. M., Hrobjartsson, A., Lalu, M. M., Li, T., Loder, E. W., Mayo-Wilson, E., McDonald, S., McGuinness, ... & Moher, D. (2021). The PRISMA 2020 statement: an updated guideline for reporting systematic reviews. *International Journal of Surgery, 88*, 105906. https://doi.org/10.1016/j.ijsu.2021.105906 Science Direct

Project Management Institute (PMI), (2021). *The Standard for Project Management and A Guide to the Project Management Body of Knowledge (PMBOK® Guide)* (7th ed.). PMI.

Rawat, A., Gupta, S., & Rao, T. (2022). A review on prospective risks and mitigation for oil and gas projects: implication for Indian CGD companies. *International Journal of Energy Sector Management, 17*(1), 41-62. https://doi.org/10.1108/ijesm-01-2021-0016 Emerald Insight

Roshdi, F., Ismail, K., Lop, N., & Wahab, L. (2022). Cost Overruns in Engineering Procurement Construction (EPC) Fabrication Oil and Gas Megaprojects in Malaysia: The Importance of Resource Allocation (5M). *International Journal of Academic Research in Business and Social Sciences, 12*(9), 1491-1499. https://doi.org/10.6007/ijarbss/v12-i9/14752 Google Scholar

Saleh, C., Hidayati, F., & Rasyid, N. (2023). Public Human Resources Development Systematic Literature Review: Competency-Based HR Development Strategic Planning. *ASSEHR 761*, 249-262. https://doi.org/10.2991/978-2-38476-082-4_24 Google Scholar

Satria, R., Arifin, R., & Sari, E. (2023). Proposed Project Selection & Prioritization Framework during Front-End Planning (FEP) for Revitalization Project Portfolio at Indonesian Oil Refinery Plant. *Jurnal Teknik Industri, 13*(1), 1-15. https://doi.org/10.25105/jti.v13i1.17510 Google Scholar

Seddeeq, A., B., Assaf, A., Sadi, A., Abdallah, A., & Hassanain, M. A. (2019). Time and Cost Overrun in the Saudi Arabian Oil and Gas Construction Industry. *Buildings, 9*(2), 41. https://doi.org/10.3390/buildings9020041 Google Scholar

 https://doi.org/10.52202/081537-0002

Senaratne, S., KC, A., & Rai, S. (2023). Stakeholder management challenges and strategies for sustainability issues in megaprojects: case studies from Australia. *Built Environment Project and Asset Management, 14*(3), 414-431. https://doi.org/10.1108/bepam-11-2022-0183 Emerald Insight

Shirazi, F., Kazemipoor, H., & Tavakkoli-Moghaddam, R. (2017). Fuzzy decision analysis for project scope change management. *Decision Science Letters, 6*, 395-406. https://doi.org/10.5267/j.dsl.2017.1.003 Google Scholar

Wiguna, I., Anwar, N., Widodo, A., & Rodhi, N. (2017). Risk management effectiveness of oil and gas pipeline construction in Java Island - Indonesia. *Iptek Journal of Proceedings Series, 3*(6), 587-592. https://doi.org/10.12962/j23546026.y2017i6.3308 Google Scholar

Zhan, Z., Chen, W., Yap, J., Samsudin, S., & Abdul-Rahman, H. (2019). Earned value analysis, implementation barriers, and maturity level in oil & gas production. *The South African Journal of Industrial Engineering, 30*(4). https://doi.org/10.7166/30-4-2030 Google Scholar

Biography

Dr. Felix Ogele has over 20 years of diverse professional experience, having served in roles such as Environmental Project Manager, Project Implementation Manager, and Church Project Manager across Nigeria, the United Kingdom, and Canada. He earned his PhD in Environmental Studies from Newcastle University in the U.K. and specializes in environmental governance and implications for biodiversity and ecosystem services. Driven by a deep passion for environmental stewardship and lifelong learning, Dr. Ogele is committed to making a positive impact on communities while addressing the effects of human activities on the environment. He is currently pursuing a Master's degree in Project Management at the College of Professional Studies, Northeastern University, Toronto Campus.

Women Leading on Site: The Future of Inclusive Construction Leadership in Nigeria and Canada

Funke Fakunle
Project Management Program, College of Professional Studies
Northeastern University – Toronto, Canada.
fakunle.f@northeastern.edu

Abstract

This paper investigates the evolving role of women in on-site construction leadership within the contexts of Nigeria and Canada, highlighting the future of inclusive leadership in this traditionally male-dominated industry. Using a comparative framework, the study draws on personal experiences as a female leader in Health, Safety, and Environmental (HSE) and project administration roles on major Nigerian projects, alongside relevant Canadian case studies and existing literature. It explores systemic barriers faced by women in construction leadership, including cultural norms, safety challenges, and organizational resistance, while showcasing effective strategies for fostering gender diversity and inclusion on construction sites. The paper further examines how inclusive leadership practices enhance team communication, safety culture, and project outcomes. Policy recommendations emphasize the integration of diversity initiatives, collaborative project delivery, and sustainable leadership models to advance equity and innovation in construction management. By contributing to the discourse on diversity, ethics, and leadership in global project environments, this research underscores the strategic importance of empowering women leaders to drive safer, more resilient, and innovative construction projects across diverse cultural and regulatory landscapes.

Keywords: inclusive leadership, women on site, construction project management, gender equity, Nigeria, Canada

https://doi.org/10.52202/081537-0003

1. Introduction

The construction industry plays a pivotal role in national development, infrastructure expansion, and economic growth (Fashina et al., 2021). Yet, despite its global societal significance, it remains one of the least gender-diverse sectors, particularly, in leadership roles on construction sites (Sunindijo & Kamardeen, 2017; Lekchiri & Kamm, 2020; Seidu et al., 2022). This disparity is more pronounced in field-based operations, where decision-making has direct consequences on safety, quality, timelines, and team dynamics (Arditi, Gluch, & Holmdahl, 2013). Despite the stark differences in economic and social development between Nigeria and Canada, women in both countries continue to encounter systemic barriers to leadership on construction sites (Akinsiku & Ajala, 2018; Asadian et al., 2024). The continued underrepresentation of women in site-level leadership roles raises pressing concerns about equity, innovation, and sustainability in construction project delivery (Lekchiri & Kamm, 2020).

Building on these concerns, the motivation for this research stems from the urgent need to challenge longstanding gender norms that hinder inclusive growth and innovation in the construction industry. As construction project environments become more complex and interdependent (shaped by rapid digitalization, sustainability demands, and global labor shortages) diverse leadership emerges not only as an ethical imperative but as a critical driver of strategic success. Women offer distinct strengths in problem-solving, risk management, and stakeholder engagement that are essential to meeting the evolving demands of modern construction projects (Asadian et al., 2024). Yet, many of these contributions remain underutilized due to deeply entrenched cultural and institutional barriers (Lekchiri & Kamm, 2020).

To contextualize this challenge, the central problem addressed by this paper is the enduring exclusion of women from site-based leadership roles in construction, particularly, in Nigeria and Canada, where policy frameworks and cultural attitudes diverge but ultimately yield a shared outcome: gender inequity in on-site decision-making (Akinsiku & Ajala, 2018; Asadian et al., 2024). In Nigeria, patriarchal norms and safety concerns often confine women to off-site or administrative positions, despite rising female enrollment in engineering and construction-related programs (Jimoh et al., 2017). In Canada, although diversity and inclusion policies are more established, implicit biases and the demanding realities of remote or rotational site work continue to hinder women's advancement into leadership roles (Haramincic, 2022).

Within this context, this research critically examines the current landscape of female leadership on construction sites in Nigeria and Canada, identifying systemic barriers and highlighting successful strategies and policy innovations that foster inclusivity. Drawing on case studies, personal experience, and relevant literature, the study explores how inclusive leadership models can reshape construction project management to be more equitable, resilient, and collaborative. The objectives are threefold: (1) to compare gender inclusion trends in the construction sectors of Nigeria and Canada; (2) to evaluate how inclusive leadership contributes to sustainability and innovation on site; and (3) to propose actionable strategies that advance female leadership in field-based project roles. In doing so, the study contributes to the evolving discourse on gender equity, leadership diversity, and systemic transformation within project-based environments. As the industry faces global challenges, from climate adaptation to workforce renewal, embracing inclusive leadership emerges not only as a matter of representation but as a catalyst for enhanced performance and

https://doi.org/10.52202/081537-0003

ethical project delivery, thus, offering timely insights for scholars, policymakers, and industry practitioners alike.

2. Literature Review: Leadership Landscape in Construction

As awareness of equity and diversity gains momentum across industries, the construction sector is beginning to address the longstanding gender gap in site-level leadership, signaling a cautious yet significant move toward inclusivity. Shaped by policy reforms, workplace dynamics, cultural traditions, and women's professional journeys, the literature captures evolving insights into inclusion in construction leadership worldwide, offering a nuanced understanding of how gender dynamics unfold in site-based roles across Nigeria and Canada.

Although gender diversity in construction is gradually gaining traction, it remains unevenly distributed and difficult to achieve. Seidu et al. (2022) note that the UK construction industry has made considerable efforts to promote diversity, yet the sector continues to grapple with stereotypes, gendered assumptions, and unequal career progression. Similarly, Bains et al. (2025) argue that while regulatory mechanisms such as gender pay gap reporting have prompted greater awareness, practical implementation of inclusive strategies remains inconsistent across project sites and firms.

In the Nigerian context, cultural and systemic barriers remain deeply entrenched. Jimoh et al. (2016) and Akinsiku and Ajala (2018) highlight how societal expectations, lack of mentorship, and exclusion from informal networks have hindered women's entry and progression in field-based construction roles. These barriers are compounded by limited access to capital and procurement opportunities, which further marginalize women-owned construction firms.
Meanwhile, Canadian initiatives such as those discussed by Haramincic (2022) show promising signs of inclusivity in remote project settings, yet similar challenges persist. Informal workplace cultures often privilege male-dominated decision-making and limit women's access to leadership trajectories. Maurer et al. (2021) emphasize that even in well-regulated environments, retention of women in engineering and construction leadership is undermined by lack of support during transitional life phases such as parenthood.

From a psychological and well-being perspective, Sunindijo and Kamardeen (2017) underscore how high stress levels and poor work-life balance in construction roles disproportionately affect women, threatening the sustainability of their careers. Likewise, Hasan and Kamardeen (2022) explore occupational health and safety challenges that deter women from engaging fully in site roles, often citing inadequate facilities and unsupportive reporting mechanisms.
The issue of managerial competencies across genders also plays a central role. Arditi et al. (2013) found no significant difference in leadership ability between male and female managers in the Swedish construction sector. However, the perception gap—driven by deeply rooted biases—often results in unequal assignment of critical leadership responsibilities on project sites. Lekchiri and Kamm (2020) further note that the U.S. construction industry's reliance on hierarchical, command-and-control leadership styles may alienate women leaders whose approaches often prioritize collaboration and emotional intelligence.

https://doi.org/10.52202/081537-0003

In response to these systemic barriers, a growing body of research explores strategies to attract, support, and retain women in site leadership roles. Pamidimukkala and Kermanshachi (2023) identify targeted mentorship, flexible career pathways, and visible role models as key interventions. Yan et al. (2024) examine Australian diversity initiatives and find that tailored training, gender-responsive procurement, and site-level advocacy are crucial to shifting workplace cultures.

Despite these efforts, Owusu-Boadi et al. (2024) argue that workforce diversity strategies often fail to penetrate day-to-day site operations due to a lack of sustained managerial commitment and integration into project delivery frameworks. Rutherford and Daniel (2024) echo this concern, calling for systemic change that embeds equity into project planning, performance metrics, and contract structures.

Together, the reviewed literature underscores common structural and cultural barriers to women's participation in construction leadership across global contexts. However, two key gaps remain. First, much of the research focuses on general industry-level diversity metrics or managerial roles, with limited emphasis on site-based leadership, where decisions on safety, scheduling, and execution are made. Second, there is a lack of comparative analysis between countries at different stages of policy and cultural transformation, such as Nigeria and Canada, which could offer transferable insights into what works, what doesn't, and why.

This paper addresses these gaps by focusing specifically on female leadership in site-based construction roles in both Nigeria and Canada, using a comparative approach. It integrates case study analysis, practitioner insights, and literature to explore how systemic barriers manifest differently across contexts and what inclusive strategies can be adapted or scaled. By doing so, the study contributes to a deeper understanding of how inclusive leadership in construction can evolve beyond policy compliance to become a catalyst for sustainable, ethical, and high-performing project delivery.

3. Barriers and Breakthroughs in Inclusive Construction Leadership

Despite increasing global discourse on gender equity, women in construction leadership, particularly on-site, continue to encounter persistent structural, cultural, and operational barriers. While these challenges surface across both developed and developing contexts, their roots and severity often differ. Yet, amid these obstacles, emerging breakthroughs signal a promising shift, offering fertile ground for optimism and strategic policy advancement.

3.1 Structural and Cultural Barriers to Women's Leadership on Construction Sites

One of the most persistent barriers to women's leadership on construction sites is the lack of structured mentorship and sponsorship opportunities. Asadian et al. (2024) highlight that women in construction often navigate their careers without formal mentoring support, which restricts their professional growth and access to leadership roles, particularly in male-dominated site environments. Similarly, Jimoh et al. (2016) observed that in Nigeria, informal mentoring networks are often exclusionary, reinforcing a cycle where few women are visible in leadership, and even fewer are supported to advance.

https://doi.org/10.52202/081537-0003

Gender-based discrimination (both overt and systemic) continues to hinder women's progression in site leadership roles. In their study of the UK construction industry, Seidu et al. (2022) found that women frequently encounter implicit bias, with leadership traits being culturally coded as masculine. This issue is compounded in Nigeria, where Akinsiku and Ajala (2018) report that traditional gender norms still cast construction as a male-only domain, discouraging women's participation in decision-making roles. Even in Canada, where diversity policies are more established, Haramincic (2022) notes that women in remote or resource-intensive project sites often face subtle exclusion, including being left out of critical communications and leadership opportunities.

Work-life balance constraints also disproportionately affect women, particularly those in site-based leadership roles that demand long hours or travel to remote locations. Sunindijo and Kamardeen (2017) identified work stress and inflexible schedules as major deterrents for women in the construction workforce. In the Nigerian context, Jimoh et al. (2016) found that a lack of family-friendly policies and support systems continues to push women away from site-based roles. Safety concerns are another major barrier. Hasan and Kamardeen (2022) emphasize that occupational health and safety frameworks in construction often fail to consider the unique needs of women, including appropriate PPE and protection against harassment. In Nigerian sites, similar challenges persist. According to Jimoh et al. (2016), unsafe work environments, both physically and psychologically contribute to the underrepresentation of women in leadership roles on construction projects.

3.2 Breakthroughs

While barriers to women's leadership in construction persist, emerging developments across both Nigeria and Canada offer grounds for cautious optimism. A notable shift begins with education: female enrollment in architecture, engineering, and construction management programs is on the rise, signaling a gradual strengthening of the talent pipeline. Asadian et al. (2024) affirm that exposure to formal training and technical education enhances women's confidence and career readiness, positioning them for leadership roles even in field-based settings.

Equally significant is the impact of women-led firms, which are redefining project success through values-driven leadership. These organizations frequently excel in stakeholder engagement, community relations, and sustainability, dimensions increasingly central to modern construction practice (Pamidimukkala & Kermanshachi, 2023). Arditi, Gluch, and Holmdahl (2013) emphasize that female managers often demonstrate collaborative and empathetic leadership styles, resulting in improved team cohesion and reduced workplace conflict.

Policy reform has also emerged as a lever for systemic change. In Canada, inclusive procurement frameworks now incentivize gender diversity in leadership, particularly for public infrastructure contracts (Haramincic, 2022). Similarly, professional bodies in Nigeria, such as the Nigerian Institute of Building, are advocating for gender-responsive project planning at the state level (Akinsiku & Ajala, 2018). These efforts are gradually reshaping access to site-based leadership roles.

 https://doi.org/10.52202/081537-0003

Private sector initiatives are reinforcing this progress. Companies across both countries are increasingly implementing zero-tolerance harassment policies, alongside targeted gender-equity and unconscious bias training for site supervisors (Sunindijo & Kamardeen, 2017). These shifts reflect a growing recognition that psychological safety is integral to retention and performance on site.

Moreover, while women in construction continue to navigate deeply embedded structural and cultural challenges, a combination of educational access, inclusive policy, advocacy, and evolving workplace norms is beginning to change the narrative. Understanding both the persistence of barriers and the momentum of breakthroughs is critical to designing actionable strategies that foster inclusive, resilient leadership in global construction. These emerging breakthroughs offer a foundation upon which real-world examples can be analyzed to understand how inclusive leadership is being implemented on the ground in diverse contexts.

4. Case Study Examples

Building on the growing global attention to gender equity, the lived realities of women in on-site construction leadership still reflect entrenched structural, cultural, and operational challenges. While these challenges surface across both developed and developing contexts, their roots and severity often differ. Yet, amid these obstacles, emerging breakthroughs signal a promising shift, offering fertile ground for optimism and strategic policy advancement.

4.1 Nigeria: Leadership in Action on the Ground

The evolving presence of women in site-based construction leadership in Nigeria is best illustrated through practical examples. My experiences as a Project HSE and Admin Manager on two major infrastructure projects (the Diageo Water Recovery Plant and the Tolaram Infrastructure Project for BHN Logistics) provide insights into how inclusive leadership can transform project outcomes and site culture.

On the Diageo Water Recovery Plant project (2020–2021), I served as the Project HSE Manager for Concord Enviro FZE, leading safety compliance on a complex water treatment facility managed in collaboration with Diageo Global Engineering UK. One of the most critical challenges was operating during the height of the COVID-19 pandemic. I developed and implemented a project-wide COVID-19 compliance management system, which achieved over 100,000 manhours without a single injury or positive case. This success was attributed to inclusive communication, culturally adaptive training, and coordinated response drills, including live fire and medical simulations that empowered all site workers to take ownership of safety protocols.

Additionally, I built a full HSE management system (HSE-MS) from the ground up, conducting 286 inspections and 51 audits, while mentoring a team of nine safety officers. My leadership emphasized trust, listening, and proactive engagement. These are qualities that enabled my team to deliver 211 site inductions and 26 bi-weekly training sessions, fostering a culture of safety that extended beyond compliance to genuine care.

https://doi.org/10.52202/081537-0003

In a subsequent role as Project HSE and Admin Manager for BHN Logistics on the Tolaram Infrastructure Development Project (2023–2024), I led the end-to-end review and approval process for over 80 Risk Assessment and Method Statement (RAMS) documents across all project phases (civil, mechanical, electrical, commissioning), ensuring consistent alignment with evolving safety protocols. This proactive approach ensured that the project recorded zero accidents during critical stages. I also developed a stakeholder communication plan that aligned safety briefings with project progress updates, which enhanced transparency and improved stakeholder confidence.
My dual role in HSE and project administration enabled me to drive sustainability and inclusivity on site. This included overseeing safety certification reviews for 50 heavy-duty equipment, ensuring compliance for 40 units and rectifying issues with the remaining 10. I established essential infrastructure (team housing, kitchen, clinic, and site offices) which improved living and working conditions, especially for underrepresented staff. Furthermore, I conducted 1,000+ refresher inductions, achieving an 80% increase in incident-free reporting, and spearheaded waste recycling initiatives that generated over ₦15 million while cutting CO_2 emissions by 20 tons/year.

These experiences demonstrate that inclusive leadership, particularly, when led by women in site-based roles, has measurable impacts on safety, sustainability, morale, and stakeholder trust. They also challenge the long-standing perception that women are unsuited to field leadership roles in Nigeria's construction sector. By combining technical competence with emotional intelligence and collaborative problem-solving, female leaders can catalyze cultural transformation on construction sites.

4.2 Canada: Inclusive Leadership in Sustainable Urban Development

In Canada, inclusive leadership is becoming a critical driver in reshaping the construction sector, especially within the context of sustainable urban development and equity-based project delivery. Women are increasingly emerging as transformative leaders, blending technical expertise with socially responsive approaches that address both workforce diversity and environmental performance.

Notable figures like Cheyanne Hammell, Special Projects Manager at PCL Construction, exemplify this leadership model. Hammell has managed complex, high-impact projects, including the CN Tower's observation level upgrade and the Lakeridge Gardens long-term care facility, completed under Infrastructure Ontario's Accelerated Build program. Her leadership approach combines collaboration, transparency, and a commitment to mentorship, especially, for women and early-career professionals (Goodman, 2024; SiteNews, 2022).

Similarly, Anya Barkan, Sustainability Manager at Multiplex Construction Canada Ltd., has advanced environmental and social governance (ESG) in Canadian construction. Under her direction, Multiplex became the first construction firm in Canada with science-based emissions reduction targets approved by the Science Based Targets initiative. Barkan's work emphasizes decarbonization, inclusive procurement, and community partnerships that align construction with sustainability and equity objectives (Canadian Green Building Council, 2024).

https://doi.org/10.52202/081537-0003

Across the industry, inclusive project strategies, such as flexible scheduling, zero-tolerance harassment policies, and procurement from women-owned and Indigenous businesses, are gaining traction. These practices have been associated with improved retention, job satisfaction, and community trust (Canada's Building Trades Unions, n.d.; Hixson, 2022).

Collectively, these examples demonstrate how female leaders in Canada are redefining construction management. Their approaches prioritize psychological safety, stakeholder engagement, and ethical decision-making, thus, proving that inclusive leadership not only improves equity outcomes but also contributes to superior project performance and sustainable urban development.

4.3 Comparative Insights: Nigeria and Canada

A comparative view of the Nigerian and Canadian construction sectors reveals distinct but converging trends in inclusive site-based leadership. In Nigeria, female leadership is emerging against a backdrop of limited institutional support, cultural bias, and fragmented regulatory enforcement. Yet, as demonstrated in the Diageo and Tolaram project examples, women leaders are leveraging inclusive practices to deliver measurable outcomes in safety, sustainability, and stakeholder trust. These successes are often grounded in adaptive leadership, on-the-ground problem-solving, and high emotional intelligence factors that fill institutional gaps with individual excellence.

In contrast, Canada's construction industry benefits from more established gender equity policies, institutional mechanisms, and cultural support. Leaders like Cheyanne Hammell and Anya Barkan operate within ecosystems that enable inclusive procurement, ESG compliance, and flexible work arrangements. These structural supports allow women to lead not in defiance of the system but through it—aligning sustainability and equity with organizational performance.
Despite their contextual differences, both settings underscore the transformative potential of women-led leadership. In Nigeria, innovation stems from necessity and personal initiative; in Canada, it emerges from systems alignment and strategic planning. What unites both experiences is a shift toward collaborative, values-driven project environments that challenge traditional hierarchies and promote a broader definition of leadership success.

Understanding these dynamics across both Global South and Global North contexts reveals that inclusive leadership is not a static model but an adaptable strategy. When applied thoughtfully, it can address unique cultural, institutional, and operational realities while advancing global goals for gender equity and sustainable development.

5. Strategic Recommendations

Advancing inclusive leadership in construction requires systemic interventions across policy, education, and organizational culture. Based on comparative insights from Nigeria and Canada, the following strategies are proposed:

https://doi.org/10.52202/081537-0003

A. Institutionalize Mentorship Networks
Develop national mentorship platforms connecting experienced female leaders with emerging professionals. Digital mentorship hubs can facilitate cross-border learning and peer support, particularly in regions with limited access to role models. Studies show that structured mentorship and role modelling are crucial for career progression in male-dominated sectors (Jimoh et al., 2016; Lekchiri & Kamm, 2020).

B. Integrate DEI into Certification and Training
Professional bodies should embed diversity, equity, and inclusion (DEI) modules into licensing, certification, and continuing education for project managers and site supervisors. Embedding DEI principles in training addresses systemic biases and supports inclusive site environments (Pamidimukkala & Kermanshachi, 2023; Sunindijo & Kamardeen, 2017).

C. Incentivize Inclusive Hiring and Procurement
Governments and funding agencies should prioritize projects that demonstrate gender-balanced leadership and engagement with women- or minority-owned subcontractors, particularly in publicly funded construction. This aligns with inclusive procurement trends seen in both Canada and international contexts (Haramincic, 2022).

D. Strengthen Academia–Industry Collaboration
Curricula in construction-related programs should be co-developed with industry partners to reflect inclusive leadership principles and prepare students for field realities. Canadian initiatives like Build Together exemplify how early exposure and targeted education can influence gender representation (Canada's Building Trades Unions, n.d.).

E. Collect and Publish Gender-Disaggregated Data
Regular reporting on female participation at all project levels, including site leadership, should be mandated. Data-driven strategies enable tracking progress and identifying policy gaps (Owusu-Boadi et al., 2024; Seidu et al., 2022).

F. Amplify Female-Led Success Stories
Recognition through awards, conferences, and media platforms can normalize women's visibility in site leadership, countering stereotypes and inspiring broader participation. In Canada, while such efforts are already underway through initiatives like 40 Under 40 and media profiles of leaders such as Cheyanne Hammell, there remains a need for broader industry-wide adoption and consistent amplification (Hixson, 2022; Goodman, 2024; SiteNews, 2022). In contrast, in Nigeria, such recognition practices are still nascent and should be strongly encouraged as a tool for changing public perception and promoting role models for aspiring female leaders in construction.

6. Implications and Conclusion

This paper has explored the evolving landscape of inclusive construction leadership by examining the barriers and breakthroughs experienced by women in site-based roles across Nigeria and Canada. Drawing on literature, policy analysis, and firsthand case studies, the study provides

 https://doi.org/10.52202/081537-0003

evidence that while the structural and cultural contexts differ, inclusive leadership contributes significantly to improved project outcomes, workforce morale, and long-term sustainability.
In Nigeria, female leaders often work against systemic limitations, ranging from entrenched gender norms to weak institutional support. Yet, through adaptive strategies and resilient leadership, they are delivering measurable improvements in project coordination, stakeholder communication, and environmental outcomes. Case examples such as the Diageo and Tolaram projects illustrate how women in site leadership roles can influence both performance metrics and cultural transformation, even in complex, resource-constrained environments.

Conversely, the Canadian context shows the impact of established frameworks that support gender equity through inclusive procurement, flexible scheduling, and environmental-social governance (ESG) integration. Leaders like Cheyanne Hammell and Anya Barkan exemplify how institutional support allows women to lead effectively at scale, thus, aligning technical performance with community impact and sustainability.
Together, these cases reinforce that inclusive leadership is not merely a social imperative but a strategic necessity. Projects led by women tend to reflect higher levels of collaboration, safety consciousness, and responsiveness to stakeholder needs, which align with the future direction of the global construction industry.

Going forward, construction sectors in both developed and developing economies would benefit from embedding inclusive leadership as a strategic priority. This includes strengthening mentorship pipelines, aligning education with industry needs, incentivizing inclusive hiring and procurement, and making female-led success stories more visible. By doing so, construction organizations can move beyond token representation toward a truly transformative leadership model that benefits the entire project ecosystem.

In conclusion, inclusive construction leadership, particularly when embodied in site-based roles, offers a pathway to greater efficiency, equity, and resilience. As the industry confronts complex challenges related to sustainability, workforce diversity, and delivery innovation, embracing the leadership potential of women is not just beneficial but essential.

References

Akinsiku, O. E., & Ajala, N. O. (2018). An investigation of barriers to females' involvement in the Nigeria construction industry. https://ir.unilag.edu.ng/handle/123456789/7399

Arditi, D., Gluch, P., & Holmdahl, M. (2013). Managerial competencies of female and male managers in the Swedish construction industry. *Construction management and economics*, *31*(9), 979-990. https://doi.org/10.1080/01446193.2013.828845

Asadian, E., Azeez, S., Leicht, R. M., & Asadi, S. (2024). Exploring the Barriers to Women in Construction and the Opportunities Presented through Lean. *Journal of Construction Engineering and Management*, *150*(5), 04024029. https://doi.org/10.1061/JCEMD4.COENG-13895

Bains, S., Daniel, E. I., Chinyio, E., Marte Gomez, J., & Adah, C. A. (2025). Exploring best practices and policies to reduce the gender gap in the UK construction industry. *Proceedings of the Institution of Civil Engineers-Management, Procurement and Law*, 1-11. https://doi.org/10.1680/jmapl.24.00030

https://doi.org/10.52202/081537-0003

Canada's Building Trades Unions. (n.d.). *BUILD TOGETHER: WOMEN OF THE BUILDING TRADES*. Accessed on May 11, 2025, Available on https://www.buildingtrades.ca/wp-content/uploads/2024/08/BUILD-TOGETHER-Women-of-the-Building-Trades.pdf

Canadian Green Building Council. (2024, December 10). Ask the Expert: Multiplex Construction Canada. *Canada Green Building Council (CAGBC)*. Accessed on May 7, 2025, Available on https://www.cagbc.org/news-resources/cagbc-news/green-building-pioneer-2022/

Fashina, A. A., Omar, M. A., Sheikh, A. A., & Fakunle, F. F. (2021). Exploring the significant factors that influence delays in construction projects in Hargeisa. *Heliyon*, *7*(4). DOI 10.1016/j.heliyon.2021.e06826

Goodman, J. (2024, January 17). PCL manager sets her sights high. *Construction Dive*. Accessed on May 11, 2025, Available on https://www.constructiondive.com/news/pcl-manager-toronto-construction/704688/

Haramincic, A. (2022). Diversity, equity and inclusive lessons from a workplace in the Canadian Arctic. In *A Field Guide to Managing Diversity, Equality and Inclusion in Organisations* (pp. 147-162). Edward Elgar Publishing. https://doi.org/10.4337/9781800379008.00023

Hasan, A., & Kamardeen, I. (2022). Occupational health and safety barriers for gender diversity in the Australian construction industry. *Journal of Construction Engineering and Management*, *148*(9), 04022100. https://doi.org/10.1061/(ASCE)CO.1943-7862.000235

Hickey, P. J., & Cui, Q. (2020). Gender diversity in US construction industry leaders. *Journal of Management in Engineering*, *36*(5), 04020069. https://doi.org/10.1061/(ASCE)ME.1943-5479.0000838

Hixson, R. (2022, December 15). Celebrating Canada's female construction leaders. *SiteNews - News, Trends, & Insights in Canadian Construction*. Accessed on June 5, 2025, Available on https://readsitenews.com/celebrating-canadas-female-construction-leaders

Jimoh, R. A., Oyewobi, L. O., Adamu, A. N., & Bajere, P. A. (2016). Women professionals' participation in the nigerian construction industry: finding voice for the voiceless. *Organization, technology & management in construction: an international journal*, *8*(1), 1429-1436. https://doi.org/10.1515/otmcj-2016-0005

Lekchiri, S., & Kamm, J. D. (2020). Navigating barriers faced by women in leadership positions in the US construction industry: a retrospective on women's continued struggle in a male-dominated industry. *European Journal of Training and Development*, *44*(6/7), 575-594. https://doi.org/10.1108/EJTD-11-2019-0186

Maurer, J. A., Choi, D., & Hur, H. (2021). Building a diverse engineering and construction industry: Public and private sector retention of women in the civil engineering workforce. *Journal of management in engineering*, *37*(4), 04021028. https://doi.org/10.1061/(ASCE)ME.1943-5479.0000913

Naoum, S. G., Harris, J., Rizzuto, J., & Egbu, C. (2020). Gender in the construction industry: Literature review and comparative survey of men's and women's perceptions in UK construction consultancies. *Journal of management in engineering*, *36*(2), 04019042. https://doi.org/10.1061/(ASCE)ME.1943-5479.0000731

Owusu-Boadi, J., Kissi, E., Abu, I. M., Owusu, C. D., Baiden, B. K., Eluerkeh, K., & Opoku Ware, S. N. (2024). Challenges of workforce diversity uptake in the construction industry: a mixed review approach. *Engineering, Construction and Architectural Management*. https://doi.org/10.1108/ECAM-02-2024-0216

https://doi.org/10.52202/081537-0003

Pamidimukkala, A., & Kermanshachi, S. (2023). Occupational challenges of women in construction industry: Development of overcoming strategies using Delphi technique. *Journal of Legal Affairs and Dispute Resolution in Engineering and Construction*, *15*(1), 04522028. https://doi.org/10.1061/(ASCE)LA.1943-4170.0000571

Rutherford, N. N. S., & Daniel, E. I. (2024). A systematic review of the barriers and attraction strategies for females in construction. *Proceedings of the Institution of Civil Engineers-Management, Procurement and Law*, *40*(XXXX), 1-11. Accessed on May 6, 2025, Available on https://doi.org/10.1680/jmapl.24.00009

Seidu, R. D., Young, B. E., Meyer, C. L., Thorpe, J., Fong, D., & Madanayake, U. (2022, November). Gender diversity in the UK construction industry. In *IOP Conference Series: Earth and Environmental Science* (Vol. 1101, No. 3, p. 032032). IOP Publishing. DOI https://doi.org/10.1088/1755-1315/1101/3/032032

SiteNews. (2022). *Cheyanne Hammell | 40 Under 40 in Canadian Construction Winner*. Accessed on May 7, 2025, Available on https://readsitenews.com/40-under-40-winner/cheyanne-hammell/

Sunindijo, R. Y., & Kamardeen, I. (2017). Work stress is a threat to gender diversity in the construction industry. *Journal of construction engineering and management*, *143*(10), 04017073. https://doi.org/10.1061/(ASCE)CO.1943-7862.0001387

Yan, D., Sunindijo, R. Y., & Wang, C. C. (2024). Analysis of gender diversity initiatives to empower women in the Australian construction industry. *Buildings*, *14*(6), 1707. https://doi.org/10.3390/buildings14061707

Biography

Funke Fakunle is a construction-focused project management professional with over 13 years of experience delivering complex infrastructure, logistics, and industrial projects across Nigeria. She is known for driving health, safety, and environmental (HSE) compliance while leading site-based teams to deliver efficient, inclusive, and high-performing outcomes. Funke has managed multi-billion-naira projects, integrating sustainability, innovation, and collaborative delivery systems to enhance project delivery.

Currently completing a master's in project management with a concentration in Construction at Northeastern University, Canada, Funke bridges academic insight with real-world execution. Her roles as Project HSE and Admin Manager on the Diageo Water Recovery Plant and Tolaram Infrastructure Project demonstrate her ability to transform challenging site conditions into opportunities for inclusive leadership and measurable project success.

She has co-authored over 16 articles in reputable journals like Heliyon and PM World Journal, with her most recent contributions titled "Exploring the Role of Communication in Canadian Construction Projects: Lessons from Failures and Successes." Her research interests span gender equity, sustainable construction, and leadership development in both Global South and North contexts.

Funke is passionate about creating safer, more inclusive construction environments through mentorship, policy advocacy, and capacity building. She also contributes to her community through volunteer roles with organizations like Daily Bread Food Bank, City cultural event and Evergreen Brick Works in Toronto.

She can be reached at funkefakunleca@gmail.com.

https://doi.org/10.52202/081537-0003

How can we learn in projects? Improving cost estimation in Dutch road projects

Pelle Limburg, Johan Ninan, Alfons van Marrewijk, Kirsten van Zalinge
TU Delft
P.R.Limburg@student.TUDelft.nl, J.Ninan@TUDelft.nl, A.vanMarrewijk@TUDelft.nl, Kvanzalinge@KWS.nl

Abstract

The construction sector remains one of the most inefficient and unsustainable industries, due in part to the chronic failure to learn from past projects. Within infrastructure contracting, learning is further hampered by project isolation, staff shortages, and commercial pressures. These conditions make it difficult to evaluate past performance and embed lessons into future cost estimation. This study's objective is to explore how learning from previous projects can improve the accuracy of cost estimation, by analyzing current practices, identifying barriers and enablers, and recommending strategies. Drawing on 23 interviews in the Dutch road sector, the research identifies time pressure, leadership priorities, and fragmented responsibilities as key barriers. Teams consistently prioritize preparing new bids over evaluating completed work, leaving little room for structured reflection. Revenue targets and skilled personnel shortages further reduce the time and capacity available for learning. While an open and collaborative team culture enables informal knowledge sharing, such exchanges rarely extend beyond individual projects. The study examines how leadership influences this dynamic, finding that while managers express support for learning, operational pressures often override reflection. Learning remains spontaneous unless visibly prioritized & structurally planned. To analyze where and how learning breaks down, the paper applies the multi-level learning framework for project-based organizations (PBO's) treating projects as temporary organizations embedded in a wider coordinating structure. Based on this lens, the study introduces a seventh learning process; "Iterating", to capture dynamic loops of reflection, adjustment, and knowledge integration during project execution. The findings offer insight into what it takes to institutionalize learning in fast-paced project environments: leadership that creates space for reflection, routines that embed feedback into project delivery, and cultural trust that enables open exchange.

Keywords: organizational learning, cross-project learning, cost estimation, road construction, organizational leadership

https://doi.org/10.52202/081537-0004

1. Introduction

Decades of research show that transport projects—roads, rails, bridges—often cost more than planned (Cantarelli, Molin, et al., 2012; Flyvbjerg et al., 2018; Hartmann & Dorée, 2015; Love et al., 2015; Molinari et al., 2023; Odeck, 2004, 2019). Improvements in planning methods have not solved this issue. Recent European studies report overruns in over 80% of projects, averaging 20–30% extra costs (Molinari et al., 2023). These persistent overruns suggest deeper organizational and behavioral problems. It is essential to define cost overruns precisely to understand their impact and address their prevalence. Dutch road projects follow this global trend, regularly exceeding initial budgets (Cantarelli, Flyvbjerg, et al., 2012). A key reason for ongoing overruns is poor learning from past projects. Construction teams often break up after completion, losing valuable knowledge. Reviews conducted post-project rarely improve future projects (Debs & Hubbard, 2023). Repeated mistakes like design errors and poor risk planning keep happening, showing a lack of learning and organizational memory (Durdyev, 2021).

Adding to this issue, the construction sector faces severe skilled personnel shortages. Over two-thirds of Dutch construction firms report workforce shortages, the highest among Dutch industries (Statistiek, 2025). According to the Economisch Instituut voor de Bouw (EIB), the construction workforce market is extremely tight, with high vacancy rates leading to delays, especially in infrastructure projects. This shortage disrupts learning, as fewer experienced workers mean less effective knowledge transfer, hindering accurate cost estimation and timely project completion (Bouw, 2024). Immediate project demands often overshadow long-term improvement efforts, further hindering accurate cost estimation. Despite decades of research on cost overruns, there remains a clear gap in understanding how organizations actually apply lessons from previous projects to improve cost estimation. While the phenomenon of overruns is well-documented, the use of cost-related knowledge from previous projects in practice is limited.

Although cost estimation and budget overruns are extensively studied in project management literature, less attention has been given to how organizational learning, particularly across project boundaries, might offer a missing link in improving estimation accuracy. Additionally, the attitudes and experiences of cost estimators, who may face pressure to underbid in the tender phase are rarely explored. Finally, practical strategies to embed learning into cost estimation remain underdeveloped, especially in the face of staffing constraints. Addressing this gap is crucial for enabling more accurate forecasting and reducing systemic overruns in construction projects. This gap results in the following main research question:

"How can learning from previous projects be strengthened to improve cost estimation accuracy of Dutch road construction projects?"

This study is situated within the debate on cross-project learning in project-based organizations, specifically in the context of cost estimation accuracy in Dutch road construction. While organizational learning provides the broader conceptual foundation, the study narrows in on how knowledge from past projects can be effectively transferred and institutionalized to improve estimating practices. In doing so, it examines the barriers, routines, and actor roles that shape learning continuity across temporary project boundaries.

https://doi.org/10.52202/081537-0004

2. Literature Review

Cross-project learning is rooted in the broader domains of organizational learning and knowledge management. Crossan et al. (1999) argue that an organization truly learns only when individuals reintegrate their insights into the system and when supportive structures exist to embed this learning. In project based organizations (PBOs), however, learning often remains confined within projects because teams are disbanded at the project's end (Newell, 2004). A key theoretical perspective for understanding the difficulties of learning from past projects is the view of projects as temporary organizations. According to Lundin and Söderholm (1995), such organizations are characterized by being time-bound, task-oriented, team-based, and transitional in nature. Packendorff (1995) further conceptualizes projects as processes of temporary organizing, where short-term execution goals often conflict with long-term learning needs. As a result, insights gained during one project are frequently lost, underutilized, or disconnected from future efforts. Table 1 summarizes key learning approaches alongside their main critiques.

Table 1, Learning from past projects approaches

Approach	Description and Key Features
Lessons learned reviews	Formal reviews at project milestones to capture actionable insights. Require follow-up to be effective (Newell, 2004).
Knowledge repositories	Centralized databases for project knowledge. Need consistent data entry and user engagement (Flyvbjerg, 2006).
Data-driven learning	Quantitative methods like reference class forecasting. Depend on quality and contextual fit of data (Flyvbjerg, 2006).
Project Management Office (PMO)	Central unit standardizing methods and sharing knowledge. Works best with strong team support (Julian, 2008).
Communities of practice	Informal peer groups that share tacit insights. Require cultural support (Wiewiora et al., 2010).
Staff rotation	Moves people across projects to share knowledge. Effective but may reduce team cohesion (Paver & Duffield, 2019).

Critiques of Cross-Project Learning Approaches

Despite widespread efforts to institutionalize learning, several commonly used mechanisms face persistent challenges. Lessons learned reviews are often conducted but subsequently archived without meaningful integration into future projects (Newell, 2004; Paver & Duffield, 2019). While knowledge repositories can be underutilized and may struggle to capture tacit knowledge (Bush & Tiwana, 2005) , they can also support innovation, particularly in interorganizational and multi-level environments like construction projects (Ninan et al., 2022). Data-driven learning approaches are frequently hindered by inconsistent data quality and difficulties in adapting insights to specific project contexts (Cai & Zhu, 2015). Project Management Offices (PMOs), though designed to coordinate learning, are sometimes seen as overly bureaucratic and disconnected from on-the-ground project realities (Pemsel & Wiewiora, 2013). Similarly, communities of practice often struggle to maintain long-term engagement and deliver structured, organization-wide impact (Probst & Borzillo, 2008). Finally, staff rotation, while intended to promote knowledge diffusion, can unintentionally disrupt productivity and diminish specialized expertise (Casad, 2012).

https://doi.org/10.52202/081537-0004

While these approaches offer valuable pathways for cross-project learning, their effectiveness depends on several enabling conditions. First, active leadership support and psychological safety are essential to encourage open reflection and critical feedback. Second, knowledge-sharing must be integrated into formal routines, instead of treated as a standalone exercise. Third, digital tools and repositories must be user-friendly and embedded in everyday workflows to overcome resistance and ensure uptake. Finally, aligning incentives, so that sharing knowledge is rewarded. Without these conditions, even well-designed learning strategies risk becoming symbolic rather than substantive. Table 2 highlights the key barriers that constrain learning from past projects and presents corresponding enablers identified in the literature.

Table 2, Barriers & Enablers for learning from past projects.

Barrier	Enabler
Project uniqueness and temporality	Standardization and comparability through consistent frameworks and cost structures (Flyvbjerg, 2006; Hartmann & Dorée, 2015)
Lack of formal knowledge capture	Implement institutionalized reviews and centralized repositories (Ahiaga-Dagbui & Smith, 2014; Eken et al., 2020)
Cultural and motivational barriers	Promote a culture of learning with incentives and open forums (Ahiaga-Dagbui & Smith, 2014; Eken et al., 2020)
Knowledge accessibility and quality	Invest in user-friendly systems and training to improve data usability (Hartmann & Dorée, 2015)

Despite these insights, a clear gap remains between theory and practice, particularly in knowledge transfer in cost estimation. The literature indicates that while tools for capturing "lessons learned" exist, project-based organizations struggle to actually reuse knowledge in new projects. The reviewed studies address what causes cost overruns and why (theoretically) but offer less on how to ensure that future projects internalize these lessons. This gap is clear in the limited improvement in forecast accuracy and repeated mistakes across projects. In practice, organizational and cultural barriers impede effective knowledge transfer. Consequently, there is a disconnect between the academic frameworks that advocate cross-project learning and the on-the-ground reality where such learning is minimal. This gap in applying theory to practice forms the focus for further inquiry.

3. Methods

This study uses a qualitative case approach at a prominent Dutch road contractor. Data were collected through semi-structured interviews with 23 professionals involved in project estimating and execution. This method strikes a balance between structured guidance and openness, allowing key topics such as estimating practices, use of historical data, and cultural factors to be addressed while remaining responsive to emerging insights (Adeoye-Olatunde & Olenik, 2021). Participants were purposively sampled to reflect a range of roles (Table 3). This diversity ensured that both operational and strategic perspectives on learning were captured.

Table 3, Overview of the Interview Participants

Participant	Role	Duration
1	Cost-estimator	1h 20 min

https://doi.org/10.52202/081537-0004

2	Cost-estimator	1h
3	Cost-estimator	55 min
4	Cost-estimator	40 min
5	Tender Manager	50 min
6	Tender Manager	1h 15 min
7	Project Manager	50 min
8	Project Controller	1h 20 min
9	Project Manager	45 min
10	Operations Manager	50 min
11	Business Controller	1h 20 min
12	Operations Manager	55 min
13	Project Manager	55 min
14	Operations Manager	1h
15	Project Manager	1h 20 min
16	Project Manager	1h 10 min
17	Head of Contract Management	1h 10 min
18	Site Manager	30 min
19	Operations Manager	1h
20	Site Manager	45 min
21	Head of Project Office	1h 40 min
22	Site Manager	45 min
23	Managing Director	45 min

Interview questions were informed by the literature, with follow-up prompts designed to elicit concrete examples and reflections. For example, participants were asked "Where would you start improving knowledge sharing if you were in charge?" and "How does your organization share knowledge or lessons learned from completed projects?" All interviews were recorded and transcribed verbatim. The transcripts were analyzed using an inductive thematic coding process (Gioia, 2021). First-order codes were assigned to participants' phrases and statements, keeping close to their original language. These codes were then iteratively grouped into second-order themes that represented broader patterns.

To ensure trustworthiness, transcripts were reviewed against the audio recordings to confirm accuracy, for instance, an early first-order code such as "no time for evaluation" was later split into more specific codes like "time pressure from new tenders" and "evaluation not planned upfront", as more nuanced patterns emerged across interviews. Sampling continued until thematic saturation was reached, which occurred at approximately the twentieth interview, as no major new themes emerged after that point, despite some additional instances being identified in later interviews. Ethical research practices were upheld throughout the study: participants provided informed consent, and identifying details have been anonymized.

4. Results and Discussion

First the results from the interviews will be presented followed by the framework and recommendations.

Informal Learning Culture and Practices

Interviewees consistently acknowledged the importance of structured reflection for improving cost estimation accuracy. While most could recall only a few formal evaluations, the need for such learning moments was widely endorsed.

https://doi.org/10.52202/081537-0004

"If I count my years in execution, it's been about ten years. In that time, I've experienced just one formal project evaluation. That's it." (Interview 15)

This remark typifies a broader sentiment: although people recognize the value of evaluations, they remain rare in practice. Others echoed this, explaining that formal reviews are often forgotten once a project ends, especially when teams move quickly to new tenders.

Nevertheless, several participants emphasized that informal, mid-project feedback is often more relevant and actionable than formal end-of-project reviews. These spontaneous interactions allow lessons to be directly applied in ongoing tenders. For example, one cost estimator (Interview 2) described how he adjusted trench productivity figures in a new bid after discussing actual site performance with a site manager. This form of real-time learning, although effective, often remains undocumented. The same interviewee acknowledged that such moments of reflection typically stay confined to individual projects, lacking a structured mechanism for wider organizational learning.

Informal Open Learning Culture Within Teams

A notable theme across the interviews was the presence of an open, approachable, and collaborative team culture. Many professionals described a flat organizational structure in which knowledge sharing happens freely and without hierarchy. One cost estimator (Interview 1) explained that within the cost-estimation team, people feel equal and are comfortable approaching colleagues directly to ask questions and exchange insights. Weekly cost estimation team meetings were described as important spaces for peer learning, where colleagues often share past experiences and suggest alternative approaches to current estimation challenges.

Importantly, this openness extends beyond day-to-day collaboration. Cost estimators described an atmosphere of psychological safety, which is critical for honest learning. As one cost estimator put it:

"Openness ensures that people stay honest. If that openness wasn't there, people would stop speaking up. But because it is, and people expect it, things stay healthy. If someone doesn't behave collegially, it gets noticed." (Interview 1)

This no-blame culture supports transparency and builds the trust needed to acknowledge and learn from mistakes. Learning is also encouraged across functions. Interviewee 2 noted that cost estimators are supported in reaching out to site managers and project staff to validate or refine their assumptions, especially when similar types of work have been previously executed.

Estimating Tools and Infrastructure

Several structured tools facilitate learning from past projects, with the most prominent being the *moederbegroting* — a centralized cost library containing standardized items based on previous projects (Interview 19). While this system is technically formalized, its maintenance and use are dependent on individual initiative. Interviewees noted that the cost library is regularly updated based on new quotes and estimator feedback, often carried out informally by those using it. This practice reinforces a self-driven learning culture in which estimators actively contribute to

https://doi.org/10.52202/081537-0004

maintaining knowledge systems. Quality assurance is further supported through a basic four-eyes principle, whereby estimates are reviewed by a second colleague before submission, ensuring internal consistency even in the absence of formal review protocols.

Informal Knowledge Sharing Mechanisms

Beyond formal tools, informal routines play a crucial role in supporting knowledge sharing. Operational managers frequently guide team members toward colleagues who possess relevant experience with similar challenges, helping to sustain awareness of previously evaluated topics (Interview 14).

Cost assumptions are also openly debated in key decision-making moments, such as final tender pricing meetings. As one participant noted:

> *"You can say whatever you think — especially in the final tender pricing meeting. Everyone listens seriously, and we have open discussions."* (Interview 5)

This informal, people-based system enables teams to remain adaptive, even in the absence of fully standardized procedures. Over time, estimators become familiar with each other's expertise areas and naturally seek out colleagues with relevant knowledge when needed (Interview 2).

Turnover Targets Drive Behavior

KWS Infra operates under revenue targets set by its parent company, VolkerWessels. The Rotterdam–Zwijndrecht branch also has its own turnover target. Projects carry a percentage of overhead costs (Algemene Kosten or AK), which include expenses such as office space, lighting, etc. These additional costs are covered under the condition that there was sufficient project turnover. Furthermore, the regional director is evaluated against these targets. Similarly, regional business managers each oversee their own area and strive for high turnover, as their performance is also judged accordingly.

This tension helps explain why learning is often deprioritized: turnover targets drive daily decision-making more strongly than long-term improvement. The first instances and second-order themes in the aggregate dimension of culture & current practices can be found in Table 4.

Table 4, Aggregate Dimension: Culture & Current Practices

#	1st Order Instances	2nd Order Themes
1	Formal project evaluations are crucial to learn and optimize but happen too rarely	Shared belief in importance of evaluation
2	Learning from projects is essential to avoid falling behind	
3	Open, honest environment; cost-estimators walk freely into each other's offices to ask questions	Informal open learning culture
4	Cost estimators are encouraged to talk to the site managers	
5	Mid-project feedback provides more concrete lessons than only post-project reviews	
6	Weekly cost-estimation meetings to share experiences and advice	Structural learning

https://doi.org/10.52202/081537-0004

7	Central standard cost library used for the cost-estimation process and is updated regularly (pre-filled cost items as starting point)	Estimating tools and infrastructure
8	Every estimate checked by an experienced colleague (four-eyes principle)	Informal sharing
9	Team debates and aligns on contested assumptions (e.g., production rates) in final tender pricing meeting	

Barriers to Cross-Project Learning

Although the previous section revealed a strong willingness among professionals to reflect and learn from project experience, this ambition is frequently undermined by systemic and cultural barriers within the company.

Time and Resource Constraints Further Limited by Turnover Targets

The most frequently cited barrier to learning from previous projects was time. Project teams operate under significant pressure to meet deadlines, manage logistics, and respond to real-time execution challenges. In this high-paced environment, post-project evaluations are often perceived as secondary or non-essential. One cost estimator (Interview 2) explained that field staff often express skepticism about the utility or feasibility of evaluations, viewing them as an additional burden rather than a core project task.

This prioritization becomes even more pronounced when new tender opportunities arise. Several interviewees explained that contributing to upcoming bids is not only seen as more urgent, but also more valued by the organization. For example, one project manager (Interview 7) described how preparing tenders is perceived as a direct contribution to securing future revenue, whereas evaluations are often seen as optional or backward-looking.

A further structural constraint is the persistent shortage of key support roles, such as project controllers. According to one interviewee (Interview 8), the scarcity of qualified staff in this function means that existing employees are stretched thin, limiting their ability to support evaluation processes. For example, one respondent (Interview 23) noted that their company was operating with approximately eight vacant site manager positions at the time of the interview, increasing the workload for existing staff. Although executing fewer projects could theoretically relieve the pressure and create space for learning, commercial demands and turnover targets constrain such decisions. As one operational manager (Interview 23) remarked, downsizing project volume is not always feasible due to top-down financial expectations for branch performance.

Cultural Resistance and Negative Perceptions

In addition to time pressure, learning is hindered by cultural perceptions that frame evaluation as a blame exercise rather than a tool for improvement. This mindset creates reluctance to engage in open reflection. One participant (Interview 14) noted that evaluations are frequently treated as formalities, often stored away without follow-up or visibility. As a result, there is little organizational awareness of which evaluations have been completed or what their outcomes were.

https://doi.org/10.52202/081537-0004

Others expressed concern that evaluations risk becoming judgmental, leading to interpersonal friction instead of constructive dialogue. This association limits the psychological safety needed for open and honest conversations about what went wrong and why. One participant observed:

"You only end up with hassle about what did or didn't go well. And who did a good job and who didn't." (Interview 6)

This quote reflects a broader pattern, echoed in several interviews, where the absence of a clear learning culture makes people hesitant to speak openly. Instead of surfacing valuable insights, evaluations risk reinforcing defensiveness and disengagement.

Lack of Formal Structures and Ownership

A further barrier lies in the absence of formal mechanisms and clear ownership for learning. Many interviewees described learning as an ad hoc activity with no dedicated coordinator or process. As one cost estimator described this clearly when asked who is responsible for learning, noting that no single person is explicitly responsible for learning—rather, it is spread thinly across many individuals. This, they explained, is problematic because when responsibility is shared by all, it effectively belongs to no one (Interview 1). This absence of clear accountability creates confusion and inertia. Several others noted that when learning is not someone's explicit job, it becomes deprioritized, especially during busy project phases. In one case, a planner described how evaluations are postponed indefinitely simply because "no one really drives it."

Even when evaluations are conducted, their outcomes are rarely documented in a consistent manner. One participant (Interview 12) pointed out that while the head of the project office holds valuable knowledge, much of it remains in his own possession or in individual memory. Without structured documentation and feedback loops, lessons fail to travel beyond individual teams or projects.

Fragmented Collaboration Between Roles

Several interviewees pointed out that cross-role collaboration, especially between cost estimators and execution teams, is limited or inconsistent. Some estimators maintain strong ties to the field, while others rarely visit construction sites:

"There are also cost estimators who've never been out on site." (Interview 7)

This statement illustrates a broader trend noted by multiple participants: those without site exposure often rely on assumptions rather than lived experience. One field manager recounted that in such cases, estimators "miss important practical nuances," which can lead to unrealistic pricing or planning.

This creates a disconnect between those who plan and those who execute, limiting opportunities for mutual learning. A similar disconnect exists with project controllers. One participant (Interview 8) noted that these roles have become increasingly absent from evaluation processes in recent years. These organizational silos limit the depth of feedback and hinder the learning that could emerge from collaboratively interpreting project outcomes.

https://doi.org/10.52202/081537-0004

Neglect of Financial Learning

Financial performance receives limited attention during evaluations. The financials are often only superficially discussed in evaluations (Interview 8). This superficial treatment of cost performance raises questions about financial awareness among project managers. As one participant (Interview 19) wondered, while some site managers are well-informed about their financial standing, others appear more focused on managing daily operational tasks and may lack a clear view of their project's budget status. The first instances and second-order themes in the aggregate dimension of Barriers for learning can be found in Table 5.

Table 5, Overview of Instances about Barriers for learning

#	1st Order Instances	2nd Order Themes
10	Feeling of low priority from top management	Time and resource pressure hinders learning
11	Staff shortages in cost estimation and execution	
12	Knowledge isn't systematically shared across projects or departments	Lack of formal learning structure
13	Evaluations are seen as a burden and are often seen as superficial "checkbox" exercises	Negative association with evaluations
14	Fear of negative judgment (sharing risk)	
15	Cost-estimators without field experience stay desk-bound, hesitant to engage with execution teams	Limited estimator-execution collaboration
16	Lack of feedback loop from execution to estimating for too high estimates	
17	Uncontrolled cost bookings by support departments reduce site manager's financial control	Neglect of financial learning
18	Reliance on gut feeling over tender data leads to missed opportunities in execution	
19	Financial aspects get only brief attention in evaluations	
20	Execution team's mistakes rarely addressed in evaluations (focus is on calculation and external factors)	Organizational silos
21	Each branch has its own estimating approach; minimal sharing or standardization across branches	
22	Experiential learning sticks, paper knowledge transfer is problematic	
23	Distrust between units limits collaboration and knowledge exchange	

Strategies to Embed Learning

The interviews revealed that, while structural and cultural barriers inhibit cross-project learning, professionals are not short on ideas for how to overcome them. Across roles and regions, respondents demonstrated a clear willingness to improve learning practices and offered pragmatic strategies for embedding learning into the project cycle.

Creating the Conditions to Make Learning Possible

Beyond specific tools or processes, several interviewees emphasized that learning requires an enabling environment. Without structural time, team continuity, and visible leadership support, even strong intentions to reflect and improve will not materialize. One recurring suggestion was

https://doi.org/10.52202/081537-0004

to create deliberate overcapacity in project teams, offering a buffer that enables professionals to step back and reflect. For instance, one interviewee (Interview 22) explained that teams are often so structurally overloaded that mistakes occur simply due to a lack of time. Providing more space for individual and collective evaluation could, in their view, substantially accelerate learning.

Participants also stressed that leadership must visibly endorse learning — not just in principle but through daily routines and operational priorities. One proposed strategy was to assign a project administrator to large projects, thereby relieving site managers of some administrative burden and freeing up mental space for reflection (Interview 23). This point was clearly articulated by the managing director, who emphasized the symbolic and practical role of leadership in prioritizing learning:

"Management should emphasize and support the importance of learning.... Encourage that through the operations manager. If the organization shows that it values this, it really helps." (Interview 23)

These reflections underscore that learning depends not only on individual behavior but also on how roles, expectations, and workflows are designed and resourced across the organization.

Embedding Evaluation Early and Lightly

A frequently mentioned strategy was to reinstate project evaluations as a routine and structured activity — not as an administrative add-on, but as an integral part of project delivery. Several interviewees called for evaluations to be planned in advance, ideally at the moment a project is awarded, to avoid being sidelined by other priorities later on. There was broad agreement that evaluations had lost their central position in recent years, with one project manager remarking that they had really fallen by the wayside and needed to be re-emphasized (Interview 4). Some participants suggested including dedicated financial reflection moments as part of the evaluation process, particularly to benefit cost estimators. Several also stressed the importance of storing evaluation insights in a central, accessible location. One interviewee suggested that although evaluations are carried out, there is much room for improvement in how findings are shared, and that having a centralized database for lessons learned would be a valuable step forward (Interview 12).

Simplifying Data for Learning

While formal evaluations remain important, many professionals emphasized the value of simple, data-informed reflection. One branch manager argued that the current cost-booking system is too detailed and that what really matters is tracking high-level production figures, such as tons of asphalt per day (Interview 19).

Rather than measuring everything, interviewees advocated for selecting a few key performance indicators (KPIs) that could be tracked and discussed across projects:

"Don't try to measure everything... pick the most important things that drive progress. That should be decided during the kick-off or handover." (Interview 19)

https://doi.org/10.52202/081537-0004

Others proposed a more consistent use of post-calculation analysis to systematically compare planned and actual costs (Interview 1). A post-calculation (or *nacalculatie*) is a cost-analysis performed once a project is finished. It involves comparing the original budget or estimate with the actual costs incurred.

Strengthen Execution–Estimator Interaction

A major theme across interviews was the need to bridge the gap between cost estimators and execution teams. Many interviewees emphasized that informal conversations, not necessarily formal meetings, are often the most effective way for knowledge transfer.

"It's already incredibly useful when a cost estimator just spends a morning in the site cabin. It feels much more natural than a formal evaluation with a form. These conversations should just happen organically." (Interview 2)

Despite this, such interactions remain the exception rather than the rule. Several participants noted that time constraints are a persistent obstacle, even though brief visits or check-ins could already add value (Interview 2). Others pointed to emerging practices, such as project controllers conducting site visits to gather financial insights. One interviewee (Interview 18) expressed hope that these insights would be routinely shared with cost estimators.

To further reduce the disconnect between cost estimation and execution, some interviewees proposed a more regionalized structure. Assigning cost estimators to fixed regional teams could help speed up communication and promote stronger working relationships (Interview 23). This approach was also suggested as a way to accelerate onboarding and build organizational cohesion, for instance by rotating new employees through fixed regional teams (Interview 23). These reflections underscore that learning is not a detached activity but deeply embedded in the relational dynamics of daily project work. The first instances and second-order themes in the aggregate dimension of strategies for learning can be found in Table 6.

Table 6, Overview of Instances about Strategies for learning

#	1st Order Instances	2nd Order Themes
24	Make project evaluations a standard step, scheduled at the start of each project to ensure follow-through.	Institutionalize evaluation
25	Need for a central lessons-learned database	
26	Need for informal interim project evaluations	
27	Ensure there is real follow-up on evaluation outcomes, in a culture of learning rather than blaming	
28	Proper cost coding is a prerequisite for meaningful evaluations	
29	Management should visibly prioritize learning by deliberately creating overcapacity and providing resources for learning.	Create overcapacity
30	Do detailed post-calculation of actual vs estimated costs to find patterns (data-driven learning)	Data-driven learning improvement

https://doi.org/10.52202/081537-0004

31	Strengthen estimator–execution interaction through site visits and mandatory transfer meetings.	Strengthen execution-estimator interaction
32	Assign estimators to specific regions and project teams to strengthen their connection with execution	

Multi-Level Learning framework

A framework helps explain not just where learning occurs, but how it is transferred, or fails to be transferred, throughout the organization over time.

From 4I to 6I: Foundations of the Learning Framework

This study builds on a progression of organizational learning models. (Crossan et al., 1999) propose four learning processes—Intuiting, Interpreting, Integrating, and Institutionalizing—that connect individual, group, and organizational levels. (Jenkin, 2013) introduces a fifth process, Information Foraging, emphasizing how individuals actively seek data before meaning-making. (Wodnik et al., 2024) add Interaction and Incorporation, highlighting learning across organizational boundaries. These extensions (except Interaction) inform this study's multi-level framework, which focuses on how learning unfolds within and between temporary project teams and the wider firm.

Visualizing Learning Flows: Introducing the 7th I

The framework (Figure 1), integrates insights from the 4I, 5I, and 6I models, mapping both forward learning flows and feedback loops. To capture learning within ongoing projects, the framework introduces a seventh process: **Iterating**. This refers to internal cycles of reflection and adjustment during execution, including mid-project reviews, team debriefs, and cost-assumption updates. Unlike earlier models focused on learning between levels, Iterating emphasizes real-time adaptation. It responds directly to empirical findings that structured feedback during execution was often lacking.

Projects as Temporary Organizations

Each project is functions as a temporary organization: it operates with a high degree of autonomy, assembles its own team, delivers within a specific contractual scope, and disbands once completed (Lundin & Söderholm, 1995; Packendorff, 1995).

The Coordinating Body: The Construction Firm

In this framework, the construction company acts as the coordinating body overseeing multiple projects. The coordinating body plays a critical role in maintaining organizational memory and enabling continuous improvement in cost estimation over time.

Propositions

To clarify the contributions from this research to the existing literature, a series of propositions is presented that extend the core multi-level learning model (Figure 1). Each proposition specifies a presumed mechanism by which practices at one level (group, project, or coordinating body) influence learning outcomes at another level. These statements (P1–P7) enhance the framework's explanatory capacity and highlight key areas for empirical validation or targeted intervention. All propositions are grounded in the patterns that emerged from the interview coding and Gioia analysis:

https://doi.org/10.52202/081537-0004

P1: Group → Project, "Integration" Knowledge developed within functional groups, such as the cost estimation team, enhances project assumptions and planning when it is incorporated into new tenders and execution strategies.
P2: Project → Group, "Institutionalization" Lessons learned during project execution contribute to strengthened collective practices when they are routinely shared at the group level, for example, through project manager meetings.
P3: Project → Coordinating Body, "Incorporation" Insights from structured mid- and post-project evaluations support continuous learning and policy refinement when they are systematically transferred to the organizational level.
P4: Coordinating Body → Project, "Institutionalization" Centralized tools and guidelines, informed by previous learning, lead to more consistent application of best practices when they are embedded into new projects and tenders.
P5: Group → Coordinating Body, "Incorporation" Insights from recurring group-level activities, such as weekly cost estimation meetings, influence the development and refinement of standardized tools and procedures.
P6: Coordinating Body → Group, "Institutionalization" Synthesized lessons, data, and tools distributed by the coordinating body enable more informed decision-making at the group level when shared through structured formats like estimation meetings.
P7: Within the Project, "Iterating" Mid-project reflection and informal communication lead to continuous adaptation during project execution by prompting re-evaluation of assumptions and decisions, especially as new team members join.

https://doi.org/10.52202/081537-0004

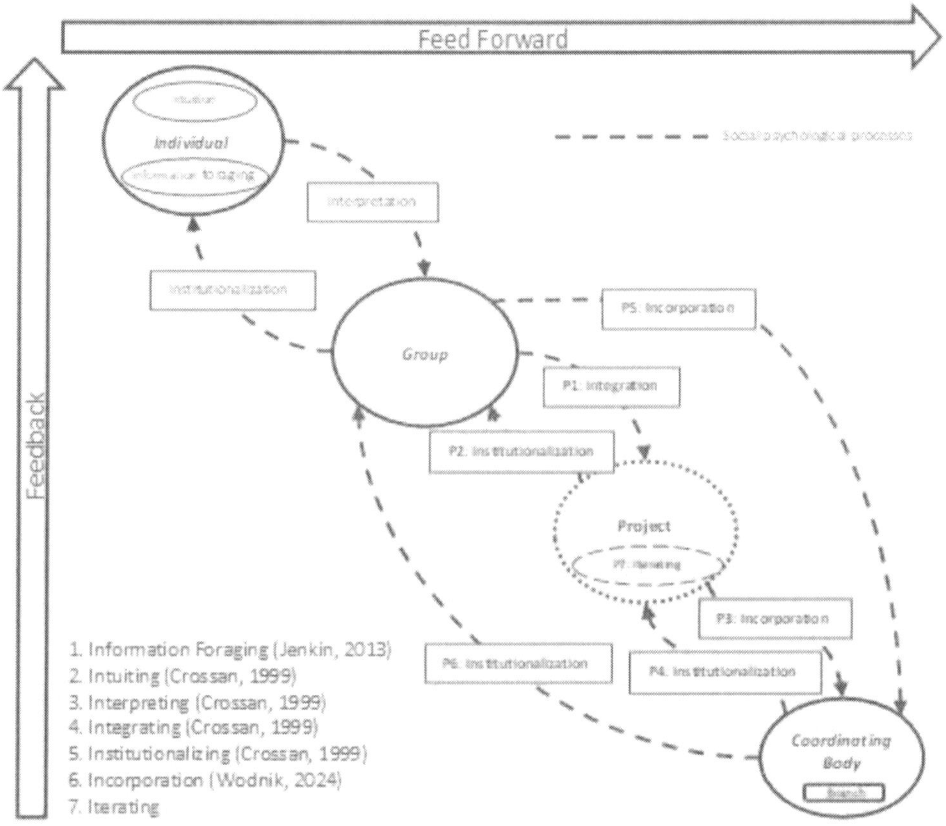

Figure 1, 7I Multi-level Framework

Individual Level – Information Foraging, Intuition, and Interpretation

Employees recognize the value of learning from past projects but are often too constrained by operational pressures to reflect or share knowledge systematically, reinforcing a persistent 'knowing–doing' gap. While many express a willingness to engage in evaluation, they lack the time, structure, and support to follow through. Low-effort routines, such as mid-project check-ins or brief handovers, can offer meaningful insight without adding burden.

However, time alone may not be sufficient. Without supportive culture, clear routines, and managerial reinforcement, extra time often gets absorbed by other priorities. Individuals may see reflection as a distraction rather than a benefit, especially under deadline pressure. This imbalance between individual investment and organizational gain helps explain why learning is frequently deprioritized.

Group Level – Integration and Interaction

At the group level, informal interactions, such as peer conversations and mentoring, play a crucial role in integrating knowledge. This aligns with the idea of communities of practice described by

https://doi.org/10.52202/081537-0004

(Wiewiora et al., 2010). Although these interactions facilitate tacit knowledge sharing, formal systems for capturing and embedding this knowledge are often weak or missing.

Project Level – Institutionalization, Iterating, and Integration
Projects function as autonomous, temporary organizations, where efforts to institutionalize learning are often disrupted by their short-lived nature (Lundin & Söderholm, 1995; Packendorff, 1995). As projects near completion, time pressure increases, leaving little room for structured evaluations or reflection before teams disband. This disbanding leads to a break in continuity and limits systematic knowledge transfer. However, because of the temporary setup and the lack of mechanisms to retain and share these lessons, insights usually stay within the project where they emerged and are not passed on. A further complication is the disconnect between financial performance and learning practices—projects may meet budget targets without reflecting on the estimation errors or near-misses that occurred, reducing the perceived urgency for post-project learning.

Coordinating Body Level – Incorporation, Institutionalization
At the coordinating body level, the organization must embed learning across projects by setting priorities, supporting shared systems, and institutionalizing routines. While tools like QHSE oversight and central databases exist, they are underused due to fragmented access, low visibility, and weak incentives.

A core challenge is the absence of mechanisms to distinguish between estimation errors caused by poor data and those resulting from deliberate underbidding. Without processes to classify these decisions, post-project reviews risk misattributing causes, which leads to flawed lessons and ineffective follow-up. Theory emphasizes the need for formalized routines and clearly defined knowledge leadership roles (Crossan et al., 1999; Julian, 2008).

To improve this, the coordinating body should adopt an iterative learning cycle. Lessons from evaluations should inform the themes and priorities of the next cycle. This approach, similar to established safety policy routines, would help transform evaluation from a retrospective formality into an adaptive organizational process.

Bridging the Temporality – Institutionalizing Learning
A major challenge is bridging the gap between project cycles. Informal insights often remain undocumented and fail to inform future practice, weakening organizational memory. This aligns with critiques by (Hartmann & Dorée, 2015) and highlights the need for stronger coordination and clear knowledge management roles.

Leadership is essential. If the management team does not visibly prioritize reflection alongside performance, learning is easily deprioritized. Reflection lacks short-term payoff, making it less visible than monthly financial targets. The framework shows that learning must occur across individual, group, project, and organizational levels. Both formal systems and informal practices are required to embed lessons across projects.

https://doi.org/10.52202/081537-0004

One of the clearest findings is that project managers play a pivotal role. Their daily actions, like scheduling reviews, enabling dialogue, or surfacing deviations, determine whether lessons are shared or lost. They translate strategic intent into practice.

Ultimately, bridging temporality requires not just structure but active leadership. The framework explains why systemic learning remains rare and supports the need to design feedback loops that reinforce knowledge use over time.

Practical recommendations

The following recommendations are structured using the Plan–Do–Check–Act (PDCA) cycle (Taylor et al., 2014) and through multiple decision-making, distinguishing between short-term (1 year) subjects such as improving staff cost data collection and long-term (3–5 years) themes such as improving overall cost booking accuracy.

Decision Matrix for Thematic Learning

To support adaptive implementation, a decision framework (Table 7) is proposed to assess whether annual interventions have achieved meaningful progress on the selected topic. These topics fall within broader themes that span a three- to five-year horizon. Only when a topic shows tangible improvement will the organization move to the next topic within the same overarching theme.

Table 7, Decision Matrix

Condition (if...)	Decision (then...)
The interventions in the annual plan were successfully implemented and adopted	Proceed to the next topic within the overarching 3–5-year theme
Only partial implementation or strong team dependency	Prioritize standardization and leadership involvement to support broader adoption
No visible improvement on the selected topic	Extend the annual focus and introduce deeper interventions (e.g., additional support or KPIs)

Balancing Immediate Project Goals and Long-term Learning

Turnover targets are fixed, and management cannot sidestep them. However, learning should not remain optional. While short-term reflection may reduce profit margins or require added capacity, these costs are strategic investments. Although reflection and evaluation may initially reduce profit margins or require additional capacity, these should be viewed as strategic investments rather than optional extras. To strike this balance, senior management must actively endorse and revisit evaluation outcomes, ensuring that follow-up is visible and aligned with broader strategic objectives. Recognizing and rewarding employee-led learning initiatives helps reinforce a culture where knowledge sharing is seen as valuable. Cross-regional learning events and exchanges among project managers can facilitate broader dissemination of insights and reduce siloed knowledge. Regular interaction between office and field staff, such as quarterly site visits, fosters mutual understanding and helps bridge operational divides. Pilot projects can be used to trial lightweight learning routines, allowing teams to scale up based on early, tangible successes. Formal mentorship roles should be established and incentivized, rewarding experienced employees who support knowledge transfer. Finally, evaluation findings should directly inform strategic planning so that recurring insights guide future tools, priorities, and thematic focus areas.

Theoretical Contribution

https://doi.org/10.52202/081537-0004

This study demonstrates that leadership plays a central role in enabling or constraining organizational learning. While prior research has acknowledged the importance of leadership support in general terms, this study shows in detail how specific leadership behaviors influence the institutionalization of learning from past projects. These behaviors include the visible prioritization of learning, consistent allocation of time and resources for learning activities, and the alignment of incentives to support knowledge sharing. The findings suggest that learning systems, such as evaluation routines or cost feedback loops, only become embedded when leaders actively support their use, follow up on outcomes, and communicate their strategic value. This advances current theory by linking organizational learning not only to formal processes, but also to the day-to-day actions and priorities demonstrated by leadership in fast-paced, delivery-focused project-based organizations.

Limitations

Role diversity in the interview participants improved triangulation and thematic consistency, however some roles, such as senior leadership, were underrepresented, which may have constrained insights into strategic decision-making. All interviewees came from a single organizational branch, limiting generalizability to other firms or sectors. The researcher's dual role as an intern may also have influenced responses. While this position facilitated access and rapport, it could have led to social desirability bias, particularly when discussing sensitive topics like blame culture or management performance.

Moreover, while the study focused on cost estimation processes, it did not systematically distinguish between inaccuracies caused by cognitive or organizational learning gaps and those resulting from deliberate strategic pricing decisions. Several interviewees noted that final tender prices are sometimes intentionally lowered to remain competitive, regardless of past project insights. This distinction between learning-related errors and strategic business behavior was outside the analytical scope of this study but may have significant implications for interpreting deviations in estimation accuracy.

Finally, the research relied on people's memories and opinions about learning from past projects, by interviewing 23 professionals a most complete image of the situation was assessed.

Recommendations for Future Research

Future studies should test interventions, like structured reviews or improved knowledge systems, across multiple projects or firms. Long-term studies could measure how estimation accuracy and learning culture develop over an extended period. Additional research should explore how staffing levels, time pressure, and workload affect an organization's capacity to support structured learning. Another important area of inquiry is understanding whether providing teams with more time leads to better learning outcomes. Studies could compare teams with time alone versus those with embedded reflection routines, supported by leadership and structure.

While patterns were consistent, future research could benefit from complementary methods such as document analysis or observational studies to validate behavioral claims. In particular, direct observation of evaluation practices or cost booking behavior could help distinguish between reported intentions and actual routines. Further inquiry should also address the difference between estimation errors from limited learning and those from strategic underbidding. Clarifying this distinction could improve the use of lessons learned and avoid misattribution in reviews.

https://doi.org/10.52202/081537-0004

Future work might explore how team composition, leadership style, and peer dynamics influence learning capacity, and test whether assigning roles such as "learning coordinators" improves estimation accuracy and knowledge transfer. Comparative studies across sectors and contract types could reveal how external context influences learning behavior and transferability of strategies.

5. Conclusion

This research concludes that improving learning requires making reflective practices a normal part of how the organization works. While estimators are motivated to improve, learning still happens on a spontaneous basis due to time pressure, missing routines, and a lack of priority. Bridging this gap between intention and action needs clear, structured efforts from the organization.

Therefore, improving cost estimation is not just a technical issue; it is a learning challenge. Organizations must institutionalize lessons learned, support them with simple tools and clear processes, and build trust between estimators and execution teams. Over time, this approach will reduce repeated mistakes and make estimates more reliable.

This research also extends organizational learning theory by treating the project not just as a site of execution, but as a temporary learning organization in its own right. While most learning models focus on individuals, groups, and permanent organizational structures, this study introduces the project as a distinct level within the learning process. By embedding the project level into the framework, the study helps explain why learning often stalls after project completion: teams disband, roles dissolve, and insights remain trapped in temporary structures.

This perspective has implications beyond the Dutch road construction sector. In any project-based environment, whether in engineering, consulting, or IT, the temporary nature of teams creates barriers. Recognizing projects as temporary organizations helps clarify where learning is likely to break down, and which interventions can improve continuity.

In conclusion, the study finds that the key to improving cost estimation accuracy in Dutch road construction lies in **prioritizing learning**. Learning must be built into the daily routines and valued by the wider organization. Only then can the industry move toward more predictable and effective project delivery.

Biography

Ing. Pelle Limburg holds a bachelor's degree in Civil Engineering from Avans University of Applied Sciences and is currently completing his master's in Construction Management and Engineering (CME) at Delft University of Technology (TU Delft). His study, conducted in collaboration with KWS Infra, investigates how learning from past infrastructure projects can enhance cost estimation practices in road construction.

Dr. Johan Ninan is Assistant Professor at the Faculty of Civil Engineering and Geosciences, Delft University of Technology (TU Delft), the Netherlands. Previously, he was Post-Doctoral Fellow at University College London (UCL). He is a guest editor of the Project Management Journal (PMJ) and is part of the international editorial board of journals such as the International Journal of Project Management (IJPM) and Construction Management and Economics (CME).

https://doi.org/10.52202/081537-0004

Dr. Alfons van Marrewijk, is Associate Professor at the Department Organization Sciences, Vrije Universiteit Amsterdam, Full Professor Construction Cultures at the Department of Management in the Built Environment at the Faculty of Architecture and the Built Environment at the Delft University of Technology, and Adjunct Professor of Project Management at BI Oslo.

Ir. Kirsten van Zalinge is Sr. Contract Manager at KWS Infra. She holds a master's in systems engineering, Policy Analysis & Management at TU Delft and has more than 25 years of experience in the field at Rijkswaterstaat and several Dutch contractors.

https://doi.org/10.52202/081537-0004

References

Adeoye-Olatunde, O. A., & Olenik, N. L. (2021). Research and scholarly methods: Semi-structured interviews. *JACCP: Journal of the American College of Clinical Pharmacy, 4*(10), 1358–1367. https://doi.org/10.1002/jac5.1441

Ahiaga-Dagbui, D. D., & Smith, S. D. (2014). Rethinking construction cost overruns: Cognition, learning and estimation. *Journal of Financial Management of Property and Construction, 19*(1), 38–54. https://doi.org/10.1108/JFMPC-06-2013-0027

Bouw, E. I. v. d. (2024). *Trends op de bouwarbeidsmarkt 2024–2028*. https://www.eib.nl/publicaties/trends-op-de-bouwarbeidsmarkt-2024-2028

Bush, A. A., & Tiwana, A. (2005). Designing sticky knowledge networks. *Communications of the ACM, 48*(5), 66–71. https://doi.org/10.1145/1060710.1060711

Cai, L., & Zhu, Y. (2015). The challenges of data quality and data quality assessment in the big data era. *Data Science Journal, 14*(0), 2. https://doi.org/10.5334/dsj-2015-002

Cantarelli, C. C., Flyvbjerg, B., & Buhl, S. L. (2012). Geographical variation in project cost performance: The Netherlands versus worldwide. *Journal of Transport Geography, 24*, 324–331. https://doi.org/10.1016/j.jtrangeo.2012.03.014

Cantarelli, C. C., Molin, E. J., van Wee, B., & Flyvbjerg, B. (2012). Characteristics of cost overruns for Dutch transport infrastructure projects and the importance of the decision to build and project phases. *Transport Policy, 22*, 49–56. https://doi.org/10.1016/j.tranpol.2012.04.001

Casad, S. (2012). Implications of job rotation literature for performance improvement practitioners. *Performance Improvement Quarterly, 25*(2), 27–41. https://doi.org/10.1002/piq.21118

Crossan, M. M., Lane, H. W., & White, R. E. (1999). An organizational learning framework: From intuition to institution. *Academy of Management Review, 24*(3), 522–537. https://doi.org/10.2307/259140

Debs, L., & Hubbard, B. (2023). Gathering and disseminating lessons learned in construction companies to support knowledge management. *Construction Economics and Building, 23*(1/2), 56–76. https://doi.org/10.5130/AJCEB.v23i1/2.8390

Durdyev, S. (2021). Review of construction journals on causes of project cost overruns. *Engineering, Construction and Architectural Management, 28*(4), 1241–1260. https://doi.org/10.1108/ecam-02-2020-0137

Eken, G., Bilgin, G., Dikmen, I., & Birgonul, M. T. (2020). A lessons-learned tool for organizational learning in construction. *Automation in Construction, 110*, 102977. https://doi.org/10.1016/j.autcon.2019.102977

Flyvbjerg, B. (2006). From Nobel Prize to project management: Getting risks right. *Project Management Journal, 37*(3), 5–15. https://doi.org/10.1177/875697280603700302

Flyvbjerg, B., Ansar, A., Budzier, A., Buhl, S., Cantarelli, C., Garbuio, M., Glenting, C., Holm, M. S., Lovallo, D., Lunn, D., Molin, E., Rønnest, A., Stewart, A., & van Wee, B. (2018). Five things you should know about cost overrun. *Transportation Research Part A: Policy and Practice, 118*, 174–190. https://doi.org/10.1016/j.tra.2018.07.013

Gioia, D. (2021). A systematic methodology for doing qualitative research. *The Journal of Applied Behavioral Science, 57*(1), 20–29. https://doi.org/10.1177/0021886320982715

Hartmann, A., & Dorée, A. (2015). Learning between projects: More than sending messages in bottles. *International Journal of Project Management, 33*(2), 341–351. https://doi.org/10.1016/j.ijproman.2014.07.006

 https://doi.org/10.52202/081537-0004

Jenkin, T. A. (2013). Extending the 4I organizational learning model: Information sources, foraging processes and tools. *Administrative Sciences, 3*(3), 96–109. https://doi.org/10.3390/admsci3030096

Julian, J. (2008). How project management office leaders facilitate cross-project learning and continuous improvement. *Project Management Journal, 39*(3), 43–58. https://doi.org/10.1002/pmj.20071

Love, P. E., Sing, C.-P., Carey, B., & Kim, J. T. (2015). Estimating construction contingency: Accommodating the potential for cost overruns in road construction projects. *Journal of Infrastructure Systems, 21*(2), 04014035. https://doi.org/10.1061/(ASCE)IS.1943-555X.0000221

Lundin, R. A., & Söderholm, A. (1995). A theory of the temporary organization. *Scandinavian Journal of Management, 11*(4), 437–455. https://doi.org/10.1016/0956-5221(95)00036-u

Molinari, L., Haezendonck, E., & Mabillard, V. (2023). Cost overruns of Belgian transport infrastructure projects: Analyzing variations over three land transport modes and two project phases. *Transport Policy, 134*, 167–179. https://doi.org/10.1016/j.tranpol.2023.02.017

Newell, S. (2004). Enhancing cross-project learning. *Engineering Management Journal, 16*(1), 12–20. https://doi.org/10.1080/10429247.2004.11415234

Ninan, J., Sergeeva, N., & Winch, G. (2022). Narrative shapes innovation: A study on multiple innovations in construction projects. *Construction Management and Economics, 40*(11–12), 884–902. https://doi.org/10.1080/01446193.2022.2037144

Odeck, J. (2004). Cost overruns in road construction—What are their sizes and determinants? *Transport Policy, 11*(1), 43–53. https://doi.org/10.1016/S0967-070X(03)00017-9

Odeck, J. (2019). Variation in cost overruns of transportation projects: An econometric meta-regression analysis of studies reported in the literature. *Transportation, 46*(4), 1345–1368. https://doi.org/10.1007/s11116-017-9836-5

Packendorff, J. (1995). Inquiring into the temporary organization: New directions for project management research. *Scandinavian Journal of Management, 11*(4), 319–333. https://doi.org/10.1016/0956-5221(95)00018-q

Paver, M., & Duffield, S. (2019). Project management lessons learned: "The elephant in the room." *The Journal of Modern Project Management, 6*(3). https://www.researchgate.net/publication/331804322_Project_management_lessons_learned_The_elephant_in_the_room

Pemsel, S., & Wiewiora, A. (2013). Project management office: A knowledge broker in project-based organisations. *International Journal of Project Management, 31*(1), 31–42. https://doi.org/10.1016/j.ijproman.2012.03.004

Probst, G., & Borzillo, S. (2008). Why communities of practice succeed and why they fail. *European Management Journal, 26*(5), 335–347. https://doi.org/10.1016/j.emj.2008.05.003

Statistiek, C. B. v. d. (2025, June 2). *Twee derde van de ondernemers ervaart personeelstekort.* https://www.cbs.nl/nl-nl/nieuws/2025/22/twee-derde-van-de-ondernemers-ervaart-personeelstekort

Taylor, M. J., McNicholas, C., Nicolay, C., Darzi, A., Bell, D., & Reed, J. E. (2014). Systematic review of the application of the plan–do–study–act method to improve quality in healthcare. *BMJ Quality & Safety, 23*(4), 290–298. https://doi.org/10.1136/bmjqs-2013-001862

Wiewiora, A., Liang, C., & Trigunarsyah, B. (2010). Inter- and intra-project knowledge transfer: Analysis of knowledge transfer techniques. *Proceedings of the PMI Research and Education Conference 2010*, 1–18. https://www.pmi.org/learning/library/project-knowledge-transfer-analysis-technique-6464

Wodnik, B. K., Andiappan, M., Di Ruggiero, E., & Lavery, J. V. (2024). The 6I model: An expanded 4I framework to conceptualise interorganisational learning in the global health sector. *BMJ Open, 14*(5), e083830. https://doi.org/10.1136/bmjopen-2023-083830

 https://doi.org/10.52202/081537-0004

Breaking Down Organizational Walls: A Systematic Review of Managerial and Organizational Barriers to Lean Construction Implementation

Amirsadegh Sedaghat[1], Hadi Jafarian[2], Adel A. Zadeh[3]
[1] University of Tehran, College of Fine Arts, School of Architecture - Tehran
sedaghat.amir2@ut.ac.ir
[2] Iran University of Science & Technology, School of Industrial Engineering - Tehran
hadi.jafarian@iust.ac.ir
[3] Northeastern University, College of Professional Studies – Toronto
a.zadeh@northeastern.edu

Abstract

The construction industry's pursuit of efficiency through lean principles faces significant challenges, particularly in the realm of organizational and managerial transformation. This study employs a systematic literature review using the PRISMA methodology to investigate managerial and organizational barriers that impede Lean Construction (LC) adoption. After analyzing 968 titles and abstracts and reviewing 89 full papers, we identified and examined 15 critical managerial and organizational barriers that significantly impact LC implementation. These barriers span from top management commitment issues to supply chain integration challenges, revealing the complex organizational dynamics that must be addressed for successful lean transformation. The study maps these barriers across the Lean Project Delivery System (LPDS) phases, demonstrating their pervasive influence throughout project lifecycles. By focusing on organizational and managerial challenges, this research provides construction companies with targeted insights for developing effective change management strategies. The findings emphasize that successful LC implementation requires fundamental organizational restructuring, enhanced stakeholder coordination, and sustained leadership commitment to overcome deeply entrenched traditional management practices.

Keywords: Lean Construction (LC), systematic review, PRISMA, sustainability, Lean Project Delivery System (LPDS)

https://doi.org/10.52202/081537-0005

1. Introduction

The evolution of production management systems has witnessed remarkable transformations since Toyota Motor Company pioneered lean production principles in the 1950s (Fujimoto, 1999; Ōno, 1988; Shingō, 1986). This revolutionary approach to eliminating waste and maximizing value has transcended manufacturing boundaries, finding applications across diverse industries seeking operational excellence (Liker, 2005; Womack & Jones, 1997). The construction sector, recognizing its potential for addressing persistent inefficiencies, began exploring lean principles in the early 1990s (Egan, 1998).

Koskela's groundbreaking work (1992) initiated the translation of lean thinking into construction contexts, establishing foundational strategies for managing projects through waste elimination and value optimization. The subsequent formalization of Lean Construction (LC) at the first International Group for LC conference marked a pivotal moment in construction management evolution (Koskela, 1993). LC represents a fundamental reimagining of construction processes, challenging traditional project delivery methods through systematic waste reduction and value enhancement (Li et al., 2017).

LC embodies the strategic adaptation of lean principles to construction's unique challenges (Sacks et al., 2010), aiming to optimize resource utilization while minimizing non-value-adding activities (Koskela et al., 2002; Pan and Pan, 2022). This innovative management philosophy confronts conventional construction approaches, fostering breakthrough improvements in productivity, quality, safety, and sustainability (Carvajal-Arango et al., 2019; Howell et al., 2017; Moaveni et al., 2019; Rincón et al., 2019). While LC's primary focus extends beyond sustainability, its principles create substantial opportunities for sustainable construction practices through waste reduction and resource optimization (Tafazzoli et al., 2020).

However, the journey toward lean transformation in construction reveals significant organizational and managerial challenges. Unlike manufacturing environments where lean principles originated, construction projects involve complex networks of stakeholders, temporary organizational structures, and deeply embedded traditional management practices. These organizational complexities create unique barriers that distinguish construction's lean journey from other industries' experiences. Understanding these managerial and organizational barriers becomes crucial for developing effective implementation strategies that acknowledge construction's distinctive organizational dynamics.

2. Research Gap

While the construction industry has increasingly embraced lean concepts over the past decade, organizations continue to encounter substantial managerial and organizational barriers that impede successful implementation (Ahmed, Hossain, and Haq 2021; Azadegan et al., 2013; Sarhan and Fox, 2013). The organizational complexity inherent in construction projects—characterized by temporary alliances, multi-layered hierarchies, and diverse stakeholder interests—creates unique challenges for lean adoption that differ markedly from manufacturing contexts where lean principles originated.

Existing research has identified various barriers to LC implementation across different economic and cultural contexts (Li et al., 2017). However, most studies have either provided broad categorizations of barriers without deep exploration of organizational dynamics or focused on technical and operational challenges while underemphasizing the critical role of organizational

https://doi.org/10.52202/081537-0005

structure and management systems in lean success. Previous investigations have catalogued barriers across multiple dimensions (Ahmed, Hossain, and Haq 2021; Albalkhy, Sweis, and Lafhaj, 2021; Bajjou and Chafi, 2018; Demirkesen et al., 2019), yet comprehensive analysis of how organizational and managerial factors specifically interact to create implementation challenges remains limited.

The literature reveals that organizational and managerial barriers often serve as root causes for other implementation challenges, suggesting their fundamental importance in lean transformation. Despite this recognition, no systematic review has exclusively examined these organizational dimensions to provide focused insights for addressing management-related implementation challenges. This gap is particularly significant given that organizational change and management commitment consistently emerge as critical success factors in lean implementations across industries.

Therefore, this study addresses the following research questions focused specifically on managerial and organizational dimensions:

Q1) What are the specific managerial and organizational barriers that impede LC implementation in the construction industry?

Q2) How do these organizational barriers manifest across different phases of the Lean Project Delivery System?

Q3) What is the relative importance and frequency of different managerial and organizational barriers based on existing research?

By concentrating on organizational and managerial barriers, this research provides targeted insights for developing organizational change strategies, leadership development programs, and management systems that support successful lean transformation in construction contexts.

3. Methods

This systematic review followed the Preferred Reporting Items for Systematic Reviews and Meta-Analyses (PRISMA) framework to identify managerial and organizational barriers to LC implementation (Moher et al., 2015). The PRISMA methodology ensures comprehensive, transparent, and reproducible research through its structured approach encompassing 27 checklist items covering all review stages, as also shown in Figure 2.

Phase 1: Planning. The research team conducted virtual brainstorming sessions to refine research questions focused on organizational barriers and establish search strategies. Keywords were selected to capture literature addressing managerial and organizational challenges in LC implementation. The final Boolean search string combined terms related to lean construction with organizational and managerial barrier concepts: ("Lean Construction" OR "lean project management" OR "lean project delivery system") AND ("barrier*" OR "challenge*" OR "obstacle*") AND ("management" OR "organizational" OR "leadership").

https://doi.org/10.52202/081537-0005

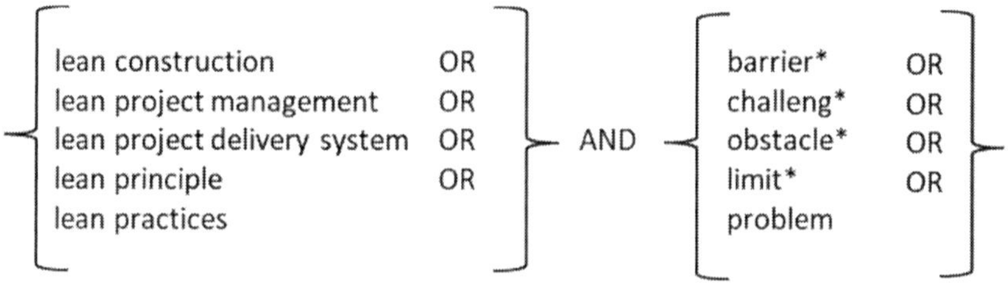

Figure 1. Boolean strings used in present SLR research

Phase 2: Development Database searches were conducted in Scopus (n=667), Web of Science (n=302), and the International Group for LC database (n=268) on July 27th, 2024. After removing duplicates using Zotero reference management software, 968 unique papers were identified. A three-member review team independently screened titles and abstracts based on inclusion criteria: (1) peer-reviewed publications, (2) construction industry focus, (3) explicit discussion of managerial/organizational barriers to LC, and (4) empirical or review methodology. This screening yielded 88 papers for full-text analysis.

Phase 3: Analysis The review team extracted data related to managerial and organizational barriers from the selected papers. Each barrier was coded, categorized, and analyzed for frequency of mention across studies. The analysis focused specifically on barriers related to organizational structure, management systems, leadership, coordination mechanisms, and organizational culture. Barriers were then mapped to LPDS phases to understand their impact across project lifecycles.

https://doi.org/10.52202/081537-0005

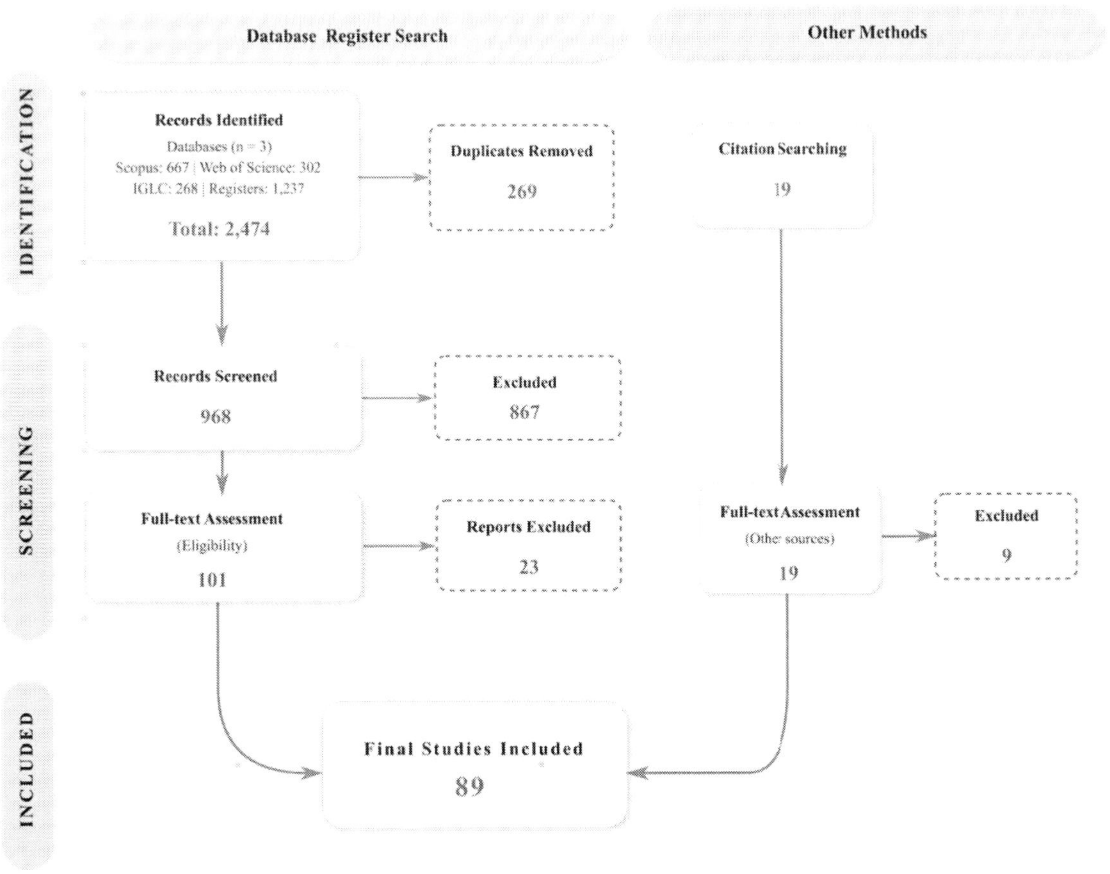

Figure 2. Flow chart of study selection process

4. Results and Discussion

4.1 Identification of Managerial and Organizational Barriers

The systematic analysis of 89 papers revealed 15 distinct managerial and organizational barriers that significantly impact LC implementation. Table 1 presents these barriers with their corresponding sources from the literature.

Table 1. Managerial and Organizational Barriers to LC Implementation

ID	Barriers	References
B1	Lack of organization and top management support and commitment in the application of LC	(1; 2; 3; 4; 5; 6; 7; 8; 9; 10; 11; 12; 13; 14; 15; 16; 18; 19; 20; 21; 22; 23; 25; 24; 17; 26; 27; 28; 29; 30; 31; 32)
B2	Limited communication among all implementation team members and the absence of mutual trust	(2; 3; 9; 10; 12; 16; 19; 20; 24; 29; 31; 32; 33; 34; 35; 36; 37; 38; 39; 40; 41; 42; 43; 44; 45; 46)

B3	Poor resource management	(2; 3; 4; 5; 10; 11; 12; 14; 19; 20; 21; 24; 27; 32; 32; 34; 36; 37; 39; 42; 43; 46; 47; 48)
B4	Lack of adoption of a systematic procurement strategy and contractual issues	(2; 3; 4; 5; 6; 12; 13; 14; 15; 16; 17; 20; 24; 26; 27; 26; 27; 35; 40; 45; 50; 51; 52; 53; 54; 55; 56)
B5	Uncertainty and lack of integration in supply chain processes, inadequate delivery performance system, and delays in material delivery	(2; 3; 4; 5; 14; 19; 20; 21; 27; 34; 35; 39; 40; 42; 56; 59)
B6	Lack of long-term philosophy, proper vision, and strategic planning	(4; 5; 10; 15; 14; 20; 27; 29; 30; 32; 40; 52; 58; 60; 61; 62; 63; 64; 67)
B7	Poor coordination among stakeholders	(2; 3; 4; 5; 6; 8; 9; 10; 12; 13; 14; 15; 16; 18; 19; 20; 21; 23; 24; 25; 27; 28; 29; 30; 31; 32; 33; 34; 35; 36; 37; 38; 39; 40; 41; 42; 43; 44; 45; 46; 50; 54; 55; 57; 58; 59; 62; 64; 65; 66; 68; 69; 70; 71; 72; 73; 74; 86; 89)
B8	Trouble with strategic alliances and long-term relationships with suppliers	(2; 3; 4; 5; 6; 11; 19; 52)
B9	Inappropriate satisfaction measurement systems and lack of focus on customer value	(2; 3; 4; 5; 18; 20; 34; 42; 43; 70; 75; 76)
B10	Lack of green initiatives	(76; 89)
B11	Management's high expectations and worries about intangible results and ambiguity in economic benefits	(1; 4; 5; 8; 9; 10; 22; 40; 49; 54; 76; 77; 85)
B12	Hierarchies in organizational structures and lengthy approval procedure from top management to take steps	(4; 5; 6; 12; 19; 20; 23; 27; 29; 38; 39; 42; 45; 50; 51; 53; 61; 63)
B13	Long implementation time needed for LC techniques application	(2; 3; 7; 10; 12; 15; 16; 17; 19; 20; 23; 28; 33; 40; 42; 50; 52; 60; 61; 62; 70; 74; 78)
B14	Centralized, delayed and poor decision making and avoiding responsibilities by top-level managers	(2; 3; 4; 5; 9; 12; 14; 15; 18; 19; 20; 21; 23; 27; 39; 43)
B15	Inadequate pre-planning and unclear project definition, improper planning and job specification	(2; 3; 4; 5; 8; 12; 16; 19; 20; 21; 30; 34; 42; 43; 49; 56; 76; 79)

1. (Ahiakwo *et al.*, 2013); 2.(Ahmed, Hossain, and Haq 2021); 3. (Ahmed and Sobuz, 2020); 4. (Albalkhy *et al.*, 2021); 5. (Albalkhy and Sweis, 2021); 6. (Aziz and Hafez, 2013); 7. (Bajjou and Chafi, 2018); 8. (Cano *et al.*, 2015); 9. (Daliri *et al.*, 2021); 10. (Demirkesen *et al.*, 2019); 11. (Elfving, 2022); 12. (Enshassi *et al.*, 2021); 13. (Evans and Farrell, 2021); 14. (Gao and Low, 2014a); 15. (Huaman-Orosco and Erazo-Rondinel, 2021); 16. (Kim and Park, 2006); 17. (Sarhan and Fox, 2012); 18. (Mano *et al.*, 2021); 19. (Moyo and Chigara, 2021); 20. (Olamilokun, 2015); 21. (Olatunji, 2008); 22. (Pan and Pan, 2022); 23 (Porwal *et al.*, 2010); 24. (Salifu-Asubay and Mensah, 2015); 25. S M et al., 2020; 26. (Sarhan and Fox, 2013); 27. (Al

https://doi.org/10.52202/081537-0005

Sayed *et al.*, 2019); 28. (Smith and Ngo, 2017); 29. (Tayeh *et al.*, 2018); 30. (Tezel *et al.*, 2018); 31. (Vignesh, 2017); 32. (Wandahl, 2014); 33. (Alarcón *et al.*, 2005); 34. (Alinaitwe, 2009); 35. (Aslam *et al.*, 2020); 36. (Fitchett and Rambuwani, 2022); 37. (Hettiaarachchige *et al.*, 2022); 38. (Khanh and Kim, 2013); 39 (Limenih *et al.*, 2022); 40. (Mossman, 2009); 41. (Nesensohn *et al.*, 2012); 42. (Sarhan *et al.*, 2018); 43. (Sholanke *et al.*, 2019); 44. (Skinnarland, 2012); 45. (Venkatesh and Venkatesan, 2021); 46. (Xing *et al.*, 2021); 47. (Anand *et al.*, 2019); 48. (Ivina and Olsson, 2020); 49. (Perez and Ghosh, 2018); 50. (Abusalem, 2020); 51. (Ahmed and Wong, 2020); 52. (Dulaimi and Tanamas, 2001); 53. (Fernandez-Solis *et al.*, 2013); 54. (Fuemana *et al.*, 2013); 55. (Johansen and Porter, 2003); 56. (Johansen and Walter, 2007); 57. (Matias and Cachadinha, 2010); 58. (Mejía-Plata *et al.*, 2016); 59. (Brady *et al.*, 2011); 60. (Alarcón *et al.*, 2002); 61. (Alsehaimi *et al.*, 2014); 62. (Bashir *et al.*, 2015); 63. (Hamzeh, 2011); 64. (Hamzeh *et al.*, 2016); 65. (Awada *et al.*, 2016); 66. (Kassab *et al.*, 2020); 67. (Torp *et al.*, 2018); 68. (Raghavan *et al.*, 2014); 69. (Ranadewa *et al.*, 2018); 70. (Ryan *et al.*, 2019); 71. (Salvatierra *et al.*, 2015); 72. (Simonsen *et al.*, 2014); 73. (Tafazzoli *et al.*, 2020); 74. (Viana *et al.*, 2010); 75. (Herrala *et al.*, 2012); 76. (Khaba and Bhar, 2017); 77. (Tezel *et al.*, 2016); 78. (Christensen *et al.*, 2019); 79. (Cerveró-Romero *et al.*, 2013); 80. (Al-Aomar, 2012); 81. (Alvarenga *et al.*, 2017); 82. (Enshassi and Zaiter, 2014); 83. (Salem *et al.*, 2005); 84. (Murguia, 2019); 85. (Tezel and Nielsen, 2013); 86. (Gao and Low, 2014b); 87. (Rooke *et al.*, 2007); 88. (Alves *et al.*, 2012); 89.(Aslam *et al.*, 2021).

4.2 Analysis of Managerial and Organizational Barriers

The 15 identified managerial and organizational barriers can be analyzed through an organizational hierarchy lens, revealing how challenges manifest differently across strategic, tactical, and operational levels. This hierarchical analysis provides insights into the multi-level nature of organizational transformation required for successful LC implementation. Understanding barriers through this framework enables targeted interventions at appropriate organizational levels, recognizing that effective lean transformation requires aligned changes across all organizational strata.

4.2.1 Strategic Level Barriers (Top Management)

Strategic level barriers represent fundamental challenges in organizational leadership, vision, and decision-making that cascade throughout the entire lean transformation process. These four barriers (B1, B6, B11, B14) collectively reflect deficiencies in executive leadership that undermine the organizational commitment necessary for sustainable change.

Lack of organization and top management support and commitment (B1) emerges as the most frequently cited barrier, appearing in 32 studies. This barrier represents more than passive disengagement; it manifests as active impediments through inadequate resource allocation, inconsistent messaging, and failure to champion cultural transformation. When executives view lean as an operational initiative rather than strategic transformation, they inadvertently signal its low priority throughout the organization. The absence of visible leadership commitment creates a vacuum where middle managers and frontline workers question the importance of lean initiatives,

https://doi.org/10.52202/081537-0005

leading to half-hearted implementation attempts that reinforce skepticism about lean's effectiveness (Wandahl, 2014; Pan and Pan, 2022; Albalkhy et al., 2021).

Lack of long-term philosophy, proper vision, and strategic planning (B6), identified in 19 studies, reveals organizations' tendencies to approach lean tactically rather than strategically. Without articulated vision linking lean principles to organizational strategy, implementation becomes a disconnected series of tool deployments lacking coherent purpose. This barrier reflects deeper issues where construction organizations, accustomed to project-based thinking, struggle to envision enterprise-wide transformation. The absence of strategic roadmaps leaves organizations vulnerable to initiative fatigue, where successive improvement attempts fail to build upon previous efforts. Strategic planning deficiencies manifest in unrealistic timelines, inadequate resource allocation, and failure to anticipate organizational resistance, ultimately undermining transformation sustainability (Torp et al., 2018; Mejía-Plata et al., 2016; Alarcón et al., 2002).

Management's high expectations and worries about intangible results and ambiguity in economic benefits (B11) creates paradoxical situations where leaders simultaneously demand transformation while doubting its value proposition. Documented in 13 studies, this barrier reflects the challenge of quantifying lean's systemic benefits within traditional financial frameworks. Construction executives, accustomed to tangible project metrics, struggle to value improvements in collaboration, learning capabilities, and problem-solving culture. This skepticism becomes self-fulfilling when leaders withdraw support before initiatives achieve critical mass, citing insufficient return on investment. The focus on immediate financial returns prevents investment in capability development and cultural change that generate long-term competitive advantages (Tezel and Nielsen, 2013; Daliri et al., 2021; Elfving, 2022).

Centralized, delayed and poor decision making and avoiding responsibilities by top-level managers (B14) represents structural impediments to Lean's emphasis on distributed leadership and rapid problem-solving. Appearing in 16 studies, this barrier reveals how traditional command-and-control structures contradict lean principles of empowered teams and continuous improvement. Centralized decision-making creates bottlenecks where operational improvements await executive approval, destroying momentum and frustrating improvement teams. Furthermore, when senior managers avoid accountability for transformation outcomes, they model behaviors antithetical to lean's emphasis on ownership and continuous learning. This barrier particularly impacts organizations attempting to implement collaborative practices like integrated project delivery, where rapid, consensus-based decisions are essential (Mano et al., 2021; Evans and Farrell, 2021; Olatunji, 2008).

4.2.2 Tactical Level Barriers (Middle Management)

Tactical level barriers represent implementation challenges at the middle management layer, where strategic vision must translate into operational reality. These four barriers (B3, B4, B12, B15) reflect the critical role of middle management in either enabling or obstructing lean transformation through their control over resources, processes, and organizational structures.

Poor resource management (B3), documented in 24 studies, encompasses both human and physical resource allocation challenges that impede lean implementation. Middle managers, caught between strategic mandates and operational pressures, often struggle to balance resources

 https://doi.org/10.52202/081537-0005

between maintaining current operations and investing in transformation initiatives. This barrier manifests in insufficient allocation of skilled personnel to improvement projects, inadequate training budgets, and competition between lean initiatives and project deadlines. The absence of systematic resource planning for lean transformation creates feast-or-famine cycles where initiatives receive intensive support followed by abandonment when project pressures mount. Human resource challenges prove particularly acute, as lean implementation requires dedicated facilitators and trained practitioners that organizations frequently fail to develop systematically (Albalkhy and Sweis, 2021; Alinaitwe, 2009; Hettiaarachchige et al., 2022).

Lack of adoption of a systematic procurement strategy and contractual issues (B4) represents institutional barriers where traditional procurement approaches contradict lean's collaborative principles. Identified in 27 studies, this barrier reflects middle management's role in perpetuating transactional relationships that impede value optimization. Lowest-bid procurement strategies, while appearing to minimize costs, actually increase total project costs through adversarial relationships, limited innovation, and extensive rework. Middle managers, evaluated on budget compliance rather than value creation, lack incentives to pioneer collaborative procurement approaches. Traditional contracts that allocate risks to individual parties create defensive behaviors incompatible with lean's emphasis on collective problem-solving and continuous improvement. This barrier particularly impacts efforts to implement integrated project delivery and early contractor involvement (Fernandez-Solis et al., 2013; Johansen and Porter, 2003; Ahmed and Wong, 2020).

Hierarchies in organizational structures and lengthy approval procedures (B12) create systematic impediments to lean's principles of rapid improvement and distributed decision-making. Noted in 18 studies, this barrier reveals how middle management layers can ossify into bureaucratic obstacles rather than value-adding coordination mechanisms. Multi-level approval requirements for simple improvements discourage innovation and reinforce cultures of compliance over initiative. Middle managers, protecting their decision-making authority, may resist lean's emphasis on team empowerment, viewing it as threatening their organizational relevance. These hierarchical structures particularly impede pull planning and Last Planner System implementation, where frontline teams require authority to make commitments and adjustments (Hamzeh, 2011; Alsehaimi et al., 2014; Porwal et al., 2010).

Inadequate pre-planning and unclear project definition, improper planning and job specification (B15) reflects tactical failures in project initiation that cascade throughout execution. Identified in 18 studies, this barrier demonstrates how middle management's rush to commence work without thorough preparation contradicts lean's emphasis on waste prevention through planning. Inadequate front-end planning creates uncertainty that compounds throughout projects, preventing the predictable workflow essential for lean construction. Unclear project definitions result from insufficient stakeholder engagement and rushed feasibility assessments, leaving teams without clear value propositions to optimize. This barrier institutionalizes waste by building uncertainty into project foundations, making subsequent lean implementation efforts analogous to building on sand (Cerveró-Romero et al., 2013; Rooke et al., 2007; Cano et al., 2015).

4.2.3 Operational Level Barriers (Implementation)

https://doi.org/10.52202/081537-0005

Operational level barriers represent challenges encountered during actual lean implementation, where abstract principles must translate into concrete practices. These seven barriers (B2, B5, B7, B8, B9, B10, B13) reflect the complexity of changing established work patterns, relationships, and systems at the operational interface where value creation occurs.

Limited communication among implementation team members and absence of mutual trust (B2) fundamentally undermines lean's collaborative foundation. Identified in 26 studies, this barrier reflects construction's fragmented nature where different specialists operate in professional silos with conflicting objectives. The temporary nature of project teams exacerbates communication challenges, as participants lack time to develop the trust relationships that enable open problem-sharing and collaborative solution development. Without psychological safety to acknowledge mistakes and share learning, teams cannot engage in the continuous improvement cycles central to lean philosophy. Information hoarding replaces transparent communication when participants fear competitive disadvantage or blame for problems, directly contradicting lean principles of visual management and shared learning (Ahmed and Sobuz, 2020; Sarhan et al., 2018; Mossman, 2009).

Uncertainty and lack of integration in supply chain processes (B5) creates operational disruptions that prevent lean flow implementation. Noted in 16 studies, this barrier reflects construction's complex material and information flows across multiple organizational boundaries without integrated coordination. Unlike manufacturing's controlled environments, construction supply chains involve numerous suppliers, subcontractors, and logistics providers operating with minimal visibility into each other's operations. This fragmentation creates uncertainty in material delivery timing and quality that forces projects to maintain wasteful buffers and contingencies. The absence of integrated planning systems prevents implementation of pull scheduling and just-in-time delivery strategies, fundamental lean construction techniques (Brady et al., 2011; Johansen and Walter, 2007; Limenih et al., 2022).

Poor coordination among stakeholders (B7) represents the most pervasive operational barrier, appearing in 59 studies. Construction projects require numerous specialized organizations to coordinate activities despite different objectives, cultures, and operational rhythms. Traditional coordination through contracts and schedules proves insufficient for lean's requirement of continuous adjustment and optimization. Poor coordination manifests in clashing trades, rework from miscommunication, and idle resources waiting for predecessor activities. The absence of effective coordination mechanisms particularly impacts collaborative lean practices like pull planning sessions, where all stakeholders must synchronize their commitments and constraints. This barrier reveals fundamental needs for new coordination approaches that transcend organizational boundaries (Demirkesen et al., 2019; Gao and Low, 2014a; Kassab et al., 2020).

Trouble with strategic alliances and long-term relationships with suppliers (B8) impedes the development of improvement partnerships essential for lean evolution. Identified in 8 studies, this barrier reflects construction's project-based nature that encourages transactional rather than relational approaches. Short-term, competitively bid relationships discourage suppliers from investing in process improvements or sharing innovations that might benefit future projects. Without stable partnerships, organizations cannot develop the mutual understanding and aligned incentives that enable collaborative improvement. This barrier particularly impacts prefabrication

 https://doi.org/10.52202/081537-0005

and modularization initiatives that require suppliers to invest in new capabilities based on long-term volume commitments (Cano et al., 2015; Ahmed and Sobuz, 2020; Albalkhy et al., 2021).

Inappropriate satisfaction measurement systems and lack of focus on customer value (B9) reveals fundamental misalignments between traditional success metrics and lean's customer-centric philosophy. Noted in 12 studies, this barrier demonstrates how operational teams, measured on cost and schedule compliance, lack frameworks for understanding and optimizing value from customer perspectives. Traditional performance indicators emphasize internal efficiency rather than effectiveness in delivering customer-defined value. Without systematic mechanisms for capturing customer feedback and translating it into operational improvements, teams perpetuate assumptions about value that may diverge significantly from actual customer needs. This barrier undermines lean's fundamental principle that customers define value (Ryan et al., 2019; Herrala et al., 2012; Perez and Ghosh, 2018).

Lack of green initiatives (B10), though mentioned in only 2 studies, represents emerging operational challenges as construction faces increasing environmental pressures. The limited attention suggests organizations have not recognized synergies between lean waste elimination and environmental sustainability. Operational teams, focused on traditional wastes of time and material, miss opportunities to simultaneously improve environmental performance. This barrier reflects organizational cultures that compartmentalize lean and green initiatives rather than integrating them into unified improvement strategies. The separation prevents leveraging lean methods for environmental benefits and using sustainability goals to drive lean implementation (Khaba and Bhar, 2017; Aslam et al., 2021).

Long implementation time needed for LC techniques application (B13) creates operational tensions between lean's evolutionary nature and construction's project-based pressures. Identified in 22 studies, this barrier reflects fundamental misalignments between lean's continuous improvement philosophy requiring patient cultivation and construction's discrete project lifecycles demanding immediate results. Operational teams, assembled for specific projects, lack continuity to sustain improvement momentum across multiple projects. The extended time required for cultural change conflicts with expectations for rapid returns, leading to premature abandonment of promising initiatives. This barrier reveals needs for new organizational forms that can maintain lean capabilities and culture beyond individual projects (Christensen et al., 2019; Salem et al., 2005; Bajjou and Chafi, 2018).

4.2.4 Barrier Prioritization

Analysis of barrier frequency across the reviewed literature reveals distinct patterns in their prevalence and perceived importance. Table 2 presents the prioritization of managerial and organizational barriers based on their frequency of citation in the literature.

Table 2. Prioritization of Managerial and Organizational Barriers

Priority Level	Barrier ID	Barrier Description	Frequency	Percentage
High	B7	Poor coordination among stakeholders	59	67.0%

https://doi.org/10.52202/081537-0005

	B1	Lack of top management support and commitment	32	36.4%
	B4	Lack of systematic procurement strategy	27	30.7%
	B2	Limited communication and absence of trust	26	29.5%
	B3	Poor resource management	24	27.3%
Medium	B13	Long implementation time needed	22	25.0%
	B6	Lack of long-term philosophy and vision	19	21.6%
	B12	Hierarchical structures and lengthy approvals	18	20.5%
	B15	Inadequate pre-planning	18	20.5%
	B5	Lack of supply chain integration	16	18.2%
	B14	Centralized and poor decision making	16	18.2%
Low	B11	Management's worries about intangible results	13	14.8%
	B9	Lack of focus on customer value	12	13.6%
	B8	Trouble with strategic alliances	8	9.1%
	B10	Lack of green initiatives	2	2.3%

The systematic analysis reveals that managerial and organizational barriers form a complex, interconnected system that fundamentally challenges LC implementation. The dominance of stakeholder coordination issues (B7) across 67% of reviewed studies underscores construction's inherent organizational complexity. This finding suggests that unlike manufacturing environments where lean originated, construction's multi-organizational structure creates unique coordination challenges that traditional management approaches cannot address effectively.

The high priority cluster (B7, B1, B4, B2, B3) represents foundational organizational capabilities without which lean transformation cannot succeed. These barriers share common characteristics: they require fundamental changes in organizational structure, culture, and management philosophy rather than simple tool implementation. The prevalence of top management support (B1) as the second most cited barrier confirms that lean transformation requires leadership-driven organizational change rather than bottom-up operational improvements alone.

Medium priority barriers (B13, B6, B12, B15, B5, B14) reveal temporal and structural challenges in aligning construction's project-based nature with lean's continuous improvement philosophy. The long implementation time required (B13) reflects tensions between lean's evolutionary transformation and construction's discrete project lifecycles. These barriers suggest needs for new organizational forms that can sustain improvement momentum across temporary project organizations.

Low priority barriers (B11, B9, B8, B10), while less frequently cited, may represent emerging challenges or context-specific issues deserving attention. The limited focus on green initiatives (B10) particularly suggests untapped opportunities for integrating lean and sustainable

https://doi.org/10.52202/081537-0005

construction practices. These barriers may gain prominence as organizations mature in their lean journeys and address more fundamental challenges.

4.2.5 Linking Barriers to LPDS Phases

Mapping managerial and organizational barriers across LPDS phases reveals their pervasive influence throughout project lifecycles. Table 3 demonstrates how these barriers create cumulative challenges that compound across phases.

Table 3. Managerial and Organizational Barriers Across LPDS Phases

LPDS Stages	Barriers	Count
Project Definition	B1, B2, B6, B7, B9, B11, B12, B13, B14	9
Lean Design	B1, B2, B4, B7, B15	5
Lean Supply	B1, B2, B3, B4, B5, B7, B8	7
Lean Assembly	B1, B2, B3, B5, B7, B8, B13, B15	8
Lean Use	B6, B7, B9	3

Project Definition Phase experiences the highest concentration of managerial barriers (9 of 15), reflecting the critical importance of organizational alignment during project initiation. Leadership commitment (B1), stakeholder coordination (B7), and strategic vision (B6) prove essential for establishing lean foundations. The concentration of barriers in this phase suggests that inadequate organizational preparation during project definition cascades into implementation challenges throughout subsequent phases.

Lean Design Phase faces fewer but critical barriers centered on communication (B2), coordination (B7), and planning (B15). These barriers prevent the collaborative design processes essential for constructability and value optimization. The presence of procurement barriers (B4) in this phase highlights how traditional contractual structures impede early contractor involvement and integrated design approaches.

Lean Supply Phase encounters barriers related to resource management (B3), supply chain integration (B5), and strategic partnerships (B8). These barriers reflect organizational challenges in extending lean principles beyond immediate project teams to broader supply networks. The persistence of coordination barriers (B7) in this phase emphasizes the difficulties in achieving seamless material and information flows across organizational boundaries.

Lean Assembly Phase experiences operational manifestations of organizational barriers, with resource management (B3) and coordination (B7) creating direct impacts on construction workflow. The reappearance of planning barriers (B15) suggests that organizational deficiencies in earlier phases create compounding challenges during execution.

Lean Use Phase shows the fewest barriers, yet their presence indicates ongoing challenges in delivering long-term value. The persistence of coordination barriers (B7) even in operational

https://doi.org/10.52202/081537-0005

phases suggests fundamental needs for new organizational approaches to facility lifecycle management.

5. Conclusion

This systematic review offers comprehensive insights into the managerial and organizational barriers that hinder LC implementation. By focusing specifically on organizational dimensions, this study reveals how deeply embedded management structures, cultures, and practices create systemic challenges for lean transformation. The identification and analysis of 15 distinct barriers provide construction organizations with targeted understanding of organizational changes required for successful lean adoption.

The research contributes to LC theory by demonstrating that organizational barriers are not merely implementation challenges but fundamental misalignments between traditional construction management and lean philosophy. The prevalence of coordination, communication, and leadership barriers across all LPDS phases confirms that lean transformation requires comprehensive organizational restructuring rather than incremental operational improvements.

For practitioners, this study offers a roadmap for organizational transformation by highlighting critical areas requiring management attention. The prioritization framework enables organizations to focus resources on high-impact barriers while understanding interconnections that create systemic challenges. By addressing these organizational barriers systematically, construction companies can create environments conducive to sustained lean implementation.

The mapping of barriers to LPDS phases provides practical insights for phase-specific interventions. Organizations can anticipate and proactively address organizational challenges at each project stage, preventing the cascade effects that amplify barriers across phases. This phase-based understanding enables more effective change management strategies aligned with project lifecycle realities.

5.1 Limitations and Future Directions

This systematic review focused exclusively on English-language, peer-reviewed publications available through major academic databases. Valuable insights from non-English sources, industry reports, or emerging practices may not be captured. Additionally, the reliance on publication frequency for prioritization may not fully reflect barriers' actual impact magnitude in different contexts. Future research should investigate the dynamic interactions between organizational barriers and their combined effects on implementation success. Longitudinal studies examining how organizations successfully overcome these barriers would provide valuable implementation insights. Empirical research measuring the relative impact of different barriers across various organizational contexts would enhance the prioritization framework developed in this study.

Investigation of emerging organizational forms that better support lean implementation represents another promising direction. Research into how digital transformation and new collaborative technologies might address traditional organizational barriers could reveal innovative solutions. Additionally, exploring the integration of lean organizational principles with sustainability and

https://doi.org/10.52202/081537-0005

resilience objectives could address the underdeveloped area of green initiatives identified in this review. The development of organizational maturity models specifically addressing these managerial and organizational barriers would provide practical assessment tools for companies beginning their lean journeys. Such frameworks could help organizations diagnose their current state and develop targeted transformation strategies addressing their specific organizational challenges.

References

Abdullah, S., Razak, A. A., & Mohammad, I. S. (2009). *Towards producing best practice in the Malaysian construction industry: The barriers in implementing the lean construction approach* [Conference paper]. *International Conference on Construction Industry 2 (ICCI2)*. https://eprints.usm.my/16097/1/shardy_Abdullah.pdf

Abusalem, O. (2020). Towards last planner system implementation in Gaza Strip, Palestine. *International Journal of Construction Management, 20*(5), 367–384. https://doi.org/10.1080/15623599.2018.1484861

Ahiakwo, O., Oloke, D., Suresh, S., & Khatib, J. (2013). A case study of Last Planner System implementation in Nigeria. In *21st Annual Conference of the International Group for Lean Construction* (pp. 636–644).

Ahmed, M., & Wong, L. S. (2020). The strategies of lean planning at selected construction sites in Klang Valley, Malaysia: A structural equation modeling approach. *Jordan Journal of Civil Engineering, 14*(3), 359–375.

Ahmed, S., Hossain, Md. M., & Haq, I. (2021). Implementation of lean construction in the construction industry in Bangladesh: Awareness, benefits and challenges. *International Journal of Building Pathology and Adaptation, 39*(2), 368–406. https://doi.org/10.1108/IJBPA-04-2019-0037

Ahmed, S., & Sobuz, M. H. R. (2020). Challenges of implementing lean construction in the construction industry in Bangladesh. *Smart and Sustainable Built Environment, 9*(2), 174–207. https://doi.org/10.1108/SASBE-02-2019-0018

Al Sayed, A. Y., Seth, D., & Hamouda, A. M. S. (2019). Prioritization of lean construction barriers in Qatari context: A fuzzy AHP approach. *International Journal of Business Excellence, 19*(4), 503–531. https://doi.org/10.1504/IJBEX.2019.103456

Al-Aomar, R. (2012). Analysis of lean construction practices at Abu Dhabi construction industry. *Lean Construction Journal*, 105–121.

Alarcón, L. F., & Calder, R. (2003). A production planning support system for construction projects. In *11th Annual Conference of the International Group for Lean Construction*.

Alarcón, L. F., Diethelm, S., Rojo, O., & Calderon, R. (2005). Assessing the impacts of implementing lean construction. 13th Annual Conference of the International Group for Lean Construction, 387–393.

Alarcón, L. F., Diethelmand, S., & Rojo, O. (2002). Collaborative Implementation of Lean Planning Systems in Chilean Construction Companies. 10th Annual Conference of the International Group for Lean Construction, 541–551.

Albalkhy, W., & Sweis, R. (2021). Barriers to adopting lean construction in the construction industry: A literature review. International Journal of Lean Six Sigma, 12(2), 210–236. https://doi.org/10.1108/IJLSS-12-2018-0144

Albalkhy, W., Sweis, R., & Lafhaj, Z. (2021). Barriers to Adopting Lean Construction in the Construction Industry-The Case of Jordan. Buildings, 11(6), 210–236. https://doi.org/10.3390/buildings11060222

Alinaitwe, H. M. (2009). Prioritizing lean construction barriers in Uganda's construction industry. Journal of Construction in Developing Countries, 14(1), 15–30.

AlManei, M., Salonitis, K., & Xu, Y. (2017). Lean Implementation Frameworks: The Challenges for SMEs. Procedia CIRP, 63, 750–755. https://doi.org/10.1016/j.procir.2017.03.170

Alsehaimi, A. O., Fazenda, P. T., & Koskela, L. (2014). Improving construction management practice with the Last Planner System: A case study. Engineering, Construction and Architectural Management, 21(1), 51–64. https://doi.org/10.1108/ECAM-03-2012-0032

Alvarenga, T. W., Da Silva, E. N., & De Brito Mello, L. C. B. (2017). BIM and lean Construction: The evolution obstacle in the Brazilian civil construction industry. Engineering, Technology & Applied Science Research, 7(5), 1904–1908. https://doi.org/10.48084/etasr.1278

Alves, T. da C. L., Milberg, C., & Walsh, K. D. (2012). Exploring lean construction practice, research, and education. Engineering, Construction and Architectural Management, 19(5), 512–525. https://doi.org/10.1108/09699981211259595

Anand, T., Sachin Prabhu, P., & Nishaant, H. A. (2019). Improvement of project performance by constraint analysis and root cause analysis of last planner system. International Journal of Recent Technology and Engineering, 7(4), 228–230.

Aslam, M., Gao, Z., & Smith, G. (2020). Development of Innovative Integrated Last Planner System (ILPS). International Journal of Civil Engineering, 18(6), 701–715. https://doi.org/10.1007/s40999-020-00504-9

Aslam, M., Gao, Z., & Smith, G. (2021). Development of lean approaching sustainability tools (Last) matrix for achieving integrated lean and sustainable construction. Construction Economics and Building, 21(3), 176–197. https://doi.org/10.5130/AJCEB.V21I3.7653

Awada, M. A., Lakkis, B. S., Doughan, A. R., & Farook R Hamzeh. (2016). Influence of lean concepts on safety in the Lebanese construction industry. https://doi.org/10.13140/RG.2.1.4273.4322

Azadegan, A., Patel, P. C., Zangoueinezhad, A., & Linderman, K. (2013). The effect of environmental complexity and environmental dynamism on lean practices. Journal of Operations Management, 31(4), 193–212. https://doi.org/10.1016/j.jom.2013.03.002

Aziz, R. F., & Hafez, S. M. (2013). Applying lean thinking in construction and performance improvement. Alexandria Engineering Journal, 52(4), 679–695. https://doi.org/10.1016/j.aej.2013.04.008

Bajjou, M. S., & Chafi, A. (2018). Lean construction implementation in the Moroccan construction industry Awareness, benefits and barriers. Journal of Engineering, Design and Technology, 16(4), 533–556. https://doi.org/10.1108/JEDT-02-2018-0031

Bashir, A. M., Suresh, S., Oloke, D. A., Proverbs, D. G., & Gameson, R. (2015). Overcoming the Challenges facing Lean Construction Practice in the UK Contracting Organizations. International Journal of Architecture, Engineering and Construction, 4(1). https://doi.org/10.7492/IJAEC.2015.002

 https://doi.org/10.52202/081537-0005

Bashir, A. M., Suresh, S., Proverbs, D. G., & Gameson, R. (2010). Barriers towards the Sustainable Implementation of Lean Construction in the United Kingdom. ARCOM Doctoral Workshop.

Brady, D., Tzortopoulos, P., & Rooke, J. (2011). An examination of the barriers to last planner implementation. 19th Annual Conference of the International Group for Lean Construction.

Cano, S., Delgado, J., Botero, L., & Rubiano, O. (2015). Barriers and success factors in lean construction implementation-survey in pilot context. 23rd Annual Conference of the International Group for Lean Construction, 631–641.

Carvajal-Arango, D., Bahamón-Jaramillo, S., Aristizábal-Monsalve, P., Vásquez-Hernández, A., & Botero, L. F. B. (2019). Relationships between lean and sustainable construction: Positive impacts of lean practices over sustainability during construction phase. Journal of Cleaner Production, 234, 1322–1337. https://doi.org/10.1016/j.jclepro.2019.05.216

Cerveró-Romero, F., Napolitano, P., Reyes, E., & Teran, L. (2013). Last Planner System® and Lean Approach Process®: Experiences from implementation in mexico. 21st Annual Conference of the International Group for Lean Construction, 645–654.

Christensen, R., Greenhalgh, S., & Thomassen, A. (2019). When a business case is not enough, motivation to work with lean. Lean Construction Journal, 2019, 121–130. https://doi.org/10.24928/2019/0146

Cudney, E., & Elrod, C. (2010). Incorporating lean concepts into supply chain management. International Journal of Six Sigma and Competitive Advantage, 6, 12–30. https://doi.org/10.1504/IJSSCA.2010.034854

Daliri, S., Young, B. K., & Lædre, O. (2021). Last Planner® System on the Minnevika Bridge Project. 29th Annual Conference of the International Group for Lean Construction, 757–766. https://doi.org/10.24928/2021/0167

Demirkesen, S., Wachter, N., Oprach, S., & Haghsheno, S. (2019). Identifying barriers in lean implementation in the construction industry. 27th Annual Conference of the International Group for Lean Construction, 157–168. https://doi.org/10.24928/2019/0151

Dulaimi, M. F., & Tanamas, C. (2001). The principles and applications of lean construction in Singapore. 9th Annual Conference of the International Group for Lean Construction.

Egan, J. (1998). Rethinking Construction (the Report of the Construction Task Force). Department of Environment, Transport and Regions.

Elfving, J. A. (2022). A decade of lessons learned: Deployment of lean at a large general contractor. Construction Management and Economics, 40(7–8), 548–561. https://doi.org/10.1080/01446193.2021.1938161

Enshassi, A., Saleh, N., & Mohamed, S. (2021). Barriers to the application of lean construction techniques concerning safety improvement in construction projects. International Journal of Construction Management, 21(10), 1044–1060. https://doi.org/10.1080/15623599.2019.1602583

Enshassi, A., & Zaiter, M. A. (2014). Implementation of lean tools on safety in construction projects in palestine. 22nd Annual Conference of the International Group for Lean Construction, 1205–1218.

Eswaramoorthi, M., Kathiresan, G. R., Prasad, P. S. S., & Mohanram, P. V. (2011). A survey on lean practices in Indian machine tool industries. International Journal of Advanced Manufacturing Technology, 52(9–12), 1091–1101. https://doi.org/10.1007/s00170-010-2788-y

Evans, M., & Farrell, P. (2021). Barriers to integrating building information modelling (BIM) and lean construction practices on construction mega-projects: A Delphi study. Benchmarking: An International Journal, 28(2), 652–669. https://doi.org/10.1108/BIJ-04-2020-0169

Evans, M., & Farrell, P. (2022). A strategic framework managing challenges of integrating lean construction and integrated project delivery on construction megaprojects, towards global integrated delivery transformative initiatives in multinational organisations. In Journal Of Engineering Design and Technology. Emerald Group Publishing Ltd. https://doi.org/10.1108/JEDT-08-2021-0402

Fernandez-Solis, J. L., Porwal, V., Lavy, S., Shafaat, A., Rybkowski, Z. K., Son, K., & Lagoo, N. (2013). Survey of motivations, benefits, and implementation challenges of last planner system users. Journal of Construction Engineering and Management, 139(4), 354–360. https://doi.org/10.1061/(ASCE)CO.1943-7862.0000606

Fitchett, A., & Rambuwani, P. (2022). Waste control in South African construction projects. Transactions of the Royal Society of South Africa. https://doi.org/10.1080/0035919X.2022.2045383

Francis, A., & Thomas, A. (2020). Exploring the relationship between lean construction and environmental sustainability: A review of existing literature to decipher broader dimensions. Journal of Cleaner Production, 252, 119913. https://doi.org/10.1016/j.jclepro.2019.119913

Fuemana, J., Puolitaival, T., & Davies, K. (2013). Last planner system—A step towards improving the productivity of new zealand construction. 21st Annual Conference of the International Group for Lean Construction, 616–625.

Fujimoto, T. (1999). The evolution of a manufacturing system at Toyota. Oxford University Press.

Gao, S., & Low, S. P. (2014a). Barriers to lean implementation in the construction industry in China. Journal of Technology Management in China, 9(2), 155–173. https://doi.org/10.1108/JTMC-12-2013-0043

Gao, S., & Low, S. P. (2014b). The Last Planner System in China's construction industry—A SWOT analysis on implementation. International Journal of Project Management, 32(7), 1260–1272. https://doi.org/10.1016/j.ijproman.2014.01.002

Green, S. D., & May, S. C. (2005). Lean construction: Arenas of enactment, models of diffusion and the meaning of 'leanness.' Building Research & Information, 33(6), 498–511. https://doi.org/10.1080/09613210500285106

Hameed, M., & Naimi, S. (2023). Examining lean construction principles and implementation obstacles. International Journal on Technical and Physical Problems of Engineering, 15(2), 237–245.

Hamzeh, F. (2011). The lean journey: Implementing the Last Planner® system in construction. 19th Annual Conference of the International Group for Lean Construction, 561–572.

Hamzeh, F., Kallassy, J., Lahoud, M., & Azar, R. (2016). The first extensive implementation of lean and LPS in Lebanon: Results and reflections. 24th Annual Conference of the International Group for Lean Construction.

Hanna, A. S., Wodalski, M., & Whited, G. (2010). Applying lean techniques in delivery of transportation infrastructure projects. 18th Annual Conference of the International Group for Lean Construction, 609–619.

Herrala, M. E., Pekuri, A., & Aapaoja, A. (2012). How do you understand lean? 20th Conference of the International Group for Lean Construction.

 https://doi.org/10.52202/081537-0005

Hettiaarachchige, N., Rathnasinghe, A., Ranadewa, K. A. T. O., & Thurairajah, N. (2022). Lean Integrated Project Delivery for Construction Procurement: The Case of Sri Lanka. Buildings, 12(5). https://doi.org/10.3390/buildings12050524

Howell, G., Ballard, G., & Demirkesen, S. (2017). Why Lean Projects Are Safer. 895–901. https://doi.org/10.24928/2017/0116

Huaman-Orosco, C., & Erazo-Rondinel, A. A. (2021). An Exploratory Study of the Main Barriers to Lean Construction Implementation in Peru. 29th Annual Conference of the International Group for Lean Construction, 474–483. https://doi.org/10.24928/2021/0173

Huaman-Orosco, C., Erazo-Rondinel, A. A., & Herrera, R. F. (2022). Barriers to Adopting Lean Construction in Small and Medium-Sized Enterprises—The Case of Peru. Buildings, 12(10). Scopus. https://doi.org/10.3390/buildings12101637

Ivina, D., & Olsson, N. O. E. (2020). Lean construction principles and railway maintenance planning. 28th Annual Conference of the International Group for Lean Construction, 565–576. https://doi.org/10.24928/2020/0025

Jadhav, J. R., Mantha, S. S., & Rane, S. B. (2014). Exploring barriers in lean implementation. International Journal of Lean Six Sigma, 5(2), 122–148. https://doi.org/10.1108/IJLSS-12-2012-0014

Jayaram, J., Droge, C., & Vickery, S. K. (1999). The impact of human resource management practices on manufacturing performance. Journal of Operations Management, 18(1), 1–20.

Johansen, E., & Porter, G. (2003). An experience of introducing last planner into a UK construction project. 11th Annual Conference of the International Group for Lean Construction.

Johansen, E., & Walter, L. (2007). Lean Construction: Prospects for the German construction industry. Lean Construction Journal, 3(1), 19–32.

Kanafani, J. (2015). Barriers to the Implementation of Lean Thinking in the Construction Industry – The Case of UAE [Master Thesis]. University of Leicester.

Kassab, O. A., Young, B. K., & Laedre, O. (2020). Implementation of Last Planner® System in an infrastructure project. 28th Annual Conference of the International Group for Lean Construction, 517–528. https://doi.org/10.24928/2020/0089

Khaba, S., & Bhar, C. (2017). Modeling the key barriers to lean construction using interpretive structural modeling. Journal of Modelling in Management, 12(4), 652–670. https://doi.org/10.1108/JM2-07-2015-0052

Khanh, H. D., & Kim, S. Y. (2013). Barriers of Last Planner System: A Survey in Vietnam Construction Industry. Journal of Construction Engineering and Project Management, 3(4), 5–11. https://doi.org/10.6106/JCEPM.2013.3.4.005

Kim, D., & Park, H.-S. (2006). Innovative construction management method: Assessment of lean construction implementation. KSCE Journal of Civil Engineering, 10(6), 381–388. https://doi.org/10.1007/BF02823976

Koskela, L. (1992). Application of the New Production Philosophy to Construction [Technical Report]. Stanford University.

Koskela, L. (1993). Lean production in construction. 1st Annual Conference of the International Group for Lean Construction.

Koskela, L., Howell, G., Ballard, G., & Tommelein, I. (2002). The foundations of lean construction. In Design and Construction: Building in Value (1st ed.).

Li, S., Wu, X., Zhou, Y., & Liu, X. (2017). A study on the evaluation of implementation level of lean construction in two Chinese firms. Renewable and Sustainable Energy Reviews, 71, 846–851. https://doi.org/10.1016/j.rser.2016.12.112

https://doi.org/10.52202/081537-0005

Liker, J. K. (2005). The Toyota way: 14 management principles from the world's greatest manufacturer. McGraw-Hill.

Likita, A. J., Jelodar, M. B., Vishnupriya, V., Rotimi, J. O. B., & Vilasini, N. (2022). Lean and BIM Implementation Barriers in New Zealand Construction Practice. Buildings, 12(10). Scopus. https://doi.org/10.3390/buildings12101645

Limenih, Z. M., Demisse, B. A., & Haile, A. T. (2022). The Usefulness of Adopting the Last Planner System in the Construction Process of Addis Ababa Road Projects. Advances in Civil Engineering. https://doi.org/10.1155/2022/7846593

Mano, A. P., Gouvea da Costa, S. E., & Pinheiro de Lima, E. (2021). Criticality assessment of the barriers to Lean Construction. International Journal of Productivity and Performance Management, 70(1), 65–86. https://doi.org/10.1108/IJPPM-11-2018-0413

Matias, J., & Cachadinha, N. (2010). Evaluating the potential benefits and challenges of lc adoption in the portuguese construction industry: A survey study. 18th Annual Conference of the International Group for Lean Construction, 285–295.

Mejía-Plata, C., Guevara-Ramirez, J. S., Moncaleano-Novoa, D. F., Londoño-Acevedo, M. C., Rojas-Quintero, J. S., & Ponz-Tienda, J. L. (2016). A route map for implementing last planner® system in Bogotá, Colombia. 24th Annual Conference of the International Group for Lean Construction, 83–92.

Moaveni, S., Banihashemi, S., & Mojtahedi, M. (2019). A Conceptual Model for a Safety-Based Theory of Lean Construction. Buildings, 9(1), 23. https://doi.org/10.3390/buildings9010023

Moher, D., Shamseer, L., Clarke, M., Ghersi, D., Liberati, A., Petticrew, M., Shekelle, P., & Stewart, L. A. (2015). Preferred reporting items for systematic review and meta-analysis protocols (PRISMA-P) 2015 statement. Systematic Reviews, 4(1), 1. https://doi.org/10.1186/2046-4053-4-1

Moradi, S., & Sormunen, P. (2023a). Implementing Lean Construction: A Literature Study of Barriers, Enablers, and Implications. Buildings, 13(2). Scopus. https://doi.org/10.3390/buildings13020556

Moradi, S., & Sormunen, P. (2023b). Integrating lean construction with BIM and sustainability: A comparative study of challenges, enablers, techniques, and benefits. Construction Innovation. Scopus. https://doi.org/10.1108/CI-02-2023-0023

Mossman, A. (2009). Why isn't the UK construction industry going lean with gusto? Lean Construction Journal, 13–23.

Moyo, T., & Chigara, B. (2021). Barriers to lean construction implementation in Zimbabwe. Journal of Engineering, Design and Technology. https://doi.org/10.1108/JEDT-01-2021-0044

Murguia, D. (2019). Factors influencing the use of last planner system methods: An empirical study in Peru. 27th Annual Conference of the International Group for Lean Construction, 1457–1468. https://doi.org/10.24928/2019/0224

Nesensohn, C., Demir, S. T., & Bryde, D. J. (2012). Developing a "True north" best practice lean company with navigational compass. 20th Annual Conference of the International Group for Lean Construction.

Nikakhtar, A., Hosseini, A. A., Wong, K. Y., & Zavichi, A. (2015). Application of lean construction principles to reduce construction process waste using computer simulation: A case study. International Journal of Services and Operations Management, 20(4), 461. https://doi.org/10.1504/IJSOM.2015.068528

 https://doi.org/10.52202/081537-0005

Okere, G. (2017). Barriers and enablers of effective knowledge management: A case in the construction sector. The Electronic Journal of Knowledge Management, 15(2), 85–97.

Olamilokun, O. (2015). Investigating Facilitators and Barriers for Adopting Lean Construction Principles in the Nigerian Building Consulting Firms. International Journal of Innovative Research & Development, 4(12), 234–239.

Olatunji, J. O. (2008). Lean-in- Nigerian Construction: State, Barriers, Strategies and "Go-Togemba" Approach. In Patricia (Ed.), 16th Annual Conference of the International Group for Lean Construction (pp. 287–297). http://www.iglc.net/papers/details/596

Ōno, T. (1988). Toyota production system: Beyond large-scale production. Productivity Press.

Onwuegbuzie, A. J., Frels, R. K., & Hwang, E. (2016). Mapping Saldaña's Coding Methods onto the Literature Review Process. Journal of Educational Issues, 2(1), 130. https://doi.org/10.5296/jei.v2i1.8931

Pan, W., & Pan, M. (2022). Rethinking lean synergistically in practice for construction industry improvements. Engineering, Construction and Architectural Management. https://doi.org/10.1108/ECAM-04-2021-0346

Perez, A. M., & Ghosh, S. (2018). Barriers faced by new-adopter of Last Planner System (R): A case study. Engineering Construction and Architectural Management, 25(9), 1110–1126. https://doi.org/10.1108/ECAM-08-2017-0162

Pheng, L. S., Gao, S., & Lin, J. L. (2015). Converging early contractor involvement (ECI) and lean construction practices for productivity enhancement some preliminary findings from Singapore. International Journal of Productivity and Performance Management, 64(6), 831–852. https://doi.org/10.1108/IJPPM-02-2014-0018

Porwal, V., Fernández-Solís, J., Lavy, S., & Rybkowski, Z. K. (2010). Last planner system implementation challenges. 18th Annual Conference of the International Group for Lean Construction, 548–556.

Prabaharan, R., & Shanmugapriya, S. (2023). Identification of Critical Barriers in Implementing Lean Construction Practices in Indian Construction Industry. Iranian Journal of Science and Technology - Transactions of Civil Engineering, 47(2), 1233–1249. Scopus. https://doi.org/10.1007/s40996-022-00959-x

Raghavan, N., Kalidindi, S., Mahalingam, A., Varghese, K., & Ayesha, A. (2014). Implementing lean concepts on indian construction sites: Organisational aspects and lessons learned. 22nd Annual Conference of the International Group for Lean Construction, 1181–1190.

Ranadewa, K., Sandanayake, Y., & Siriwardena, M. (2018). Enabling lean among small and medium enterprise (sme) contractors in Sri Lanka. 26th Annual Conference of the International Group for Lean Construction, 392–401. https://doi.org/10.24928/2018/0428

Renuka, S. M., Arunkumar, S., & Umarani, C. (2020). Implementation of Lean Construction And Its Problems On Construction Projects In India. International Journal of Civil Engineering, 7(7), 11–19. https://doi.org/10.14445/23488352/IJCE-V7I7P103

Rincón, J. L., Fernández-Solís, J. L., Lavy, S., & Du, J. (2019). Effect of Autonomous Agents on Last Planner System Performance in Texas. Journal of Management in Engineering, 35(1), 04018050. https://doi.org/10.1061/(ASCE)ME.1943-5479.0000657

Rooke, J. A., Koskela, L., & Seymour, D. (2007). Producing things or production flows? Ontological assumptions in the thinking of managers and professionals in construction. Construction Management and Economics, 25(10), 1077–1085. https://doi.org/10.1080/01446190701598665

https://doi.org/10.52202/081537-0005

Ryan, M., Murphy, C., & Casey, J. (2019). Case study in the application of the Last planner® system. 27th Annual Conference of the International Group for Lean Construction, 215–226. https://doi.org/10.24928/2019/0223

Saldaña, J. (2016). The Coding manual for qualitative researchers (3rd edition). SAGE.

Salem, O., Solomon, J., Genaidy, A., & Luegring, M. (2005). Site Implementation and Assessment of Lean Construction Techniques. Lean Construction Journal, 2(2), 1–21.

Salem, O., Solomon, J., Genaidy, A. M., & Minkarah, I. A. (2006). Lean Construction: From Theory to Implementation. Journal of Management in Engineering, 22(4), 168–175. https://doi.org/10.1061/(ASCE)0742-597X(2006)22:4(168)

Salifu-Asubay, E. K., & Mensah, C. A. (2015). Improving Delivery of Construction Projects in Ghana's Cities: A Lean Construction Approach. International Journal of Sustainable Construction Engineering and Technology, 6(1), 1–15.

Salvatierra, J. L., Alarcón, L. F., López, A., & Velásquez, X. (2015). Lean diagnosis for chilean construction industry: Towards more sustainable lean practices and tools. 23rd Annual Conference of the International Group for Lean Construction, 642–651.

Sarhan, J., Xia, B., Fawzia, S., Karim, A., & Olanipekun, A. (2018). Barriers to implementing lean construction practices in the Kingdom of Saudi Arabia (KSA) construction industry. Construction Innovation, 18(2), 246–272. https://doi.org/10.1108/CI-04-2017-0033

Sarhan, S., & Fox, A. (2012). Trends and challenges to the development of a lean culture among UK construction organizations. 20th Conference of the International Group for Lean Construction.

Sarhan, S., & Fox, A. (2013). Barriers to Implementing Lean Construction in the UK Construction Industry. The Built & Human Environment Review, 6, 1–17.

Scherrer-Rathje, M., Boyle, T. A., & Deflorin, P. (2009). Lean, take two! Reflections from the second attempt at lean implementation. Business Horizons, 52(1), 79–88. https://doi.org/10.1016/j.bushor.2008.08.004

Shingō, S. (1986). Zero quality control: Source inspection and the poka-yoke system. Productivity Press.

Sholanke, A. B., Chen, S. J., Newo, A. A., & Nwabufo, C. B. (2019). Prospects and challenges of lean construction practice in the building industry in Nigeria: Architects' perspective. International Journal of Innovative Technology and Exploring Engineering, 8(8), 667–673.

Simonsen, R., Thyssen, M. H., & Sander, D. (2014). Is lean construction another fading management concept? 22nd Annual Conference of the International Group for Lean Construction, 85–96.

Sindhwani, R., Mittal, V. K., Singh, P. L., Aggarwal, A., & Gautam, N. (2019). Modelling and analysis of barriers affecting the implementation of lean green agile manufacturing system (LGAMS). Benchmarking: An International Journal, 26(2), 498–529. https://doi.org/10.1108/BIJ-09-2017-0245

Singh, A., Kumar, V., Mittal, A., & Verma, P. (2023). Identifying critical challenges to lean construction adoption. Construction Innovation. Scopus. https://doi.org/10.1108/CI-09-2022-0229

Skinnarland, S. (2012). Norwegian project managers and foremen's experiences of collaborative planning. 20th Annual Conference of the International Group for Lean Construction.

Smith, J. P., & Ngo, K. (2017). Implementation of lean practices among finishing contractors in the US. 25th Annual Conference of the International Group for Lean Construction, 421–428. https://doi.org/10.24928/2017/0182

 https://doi.org/10.52202/081537-0005

Sweis, G. J., Hiyassat, M., & Al-Hroub, F. F. (2016). Assessing lean conformance by first-grade contractors in the Jordanian construction industry. Construction Innovation, 16(4), 446–459. https://doi.org/10.1108/CI-04-2015-0024

Sweis, R., Ghalion, R., El- Mashaleh, M., Amayreh, I., & Niveen, A.-S. (2019). The Effects of Training and Motivating Employees on Improving Performance of Construction Companies The Case of Jordan. International Journal of Information, Business and Management, 11(2), 179–211.

Tafazzoli, M., Mousavi, E., & Kermanshachi, S. (2020). Opportunities and challenges of green-lean: An integrated system for sustainable construction. Sustainability, 12(11). https://doi.org/10.3390/su12114460

Tayeh, B. A., Al Hallaq, K., Al Faqawi, A. H., Alaloul, W. S., & Kim, S. Y. (2018). Success factors and barriers of last planner system implementation in the gaza strip construction industry. Open Construction and Building Technology Journal, 12(1), 389–403. https://doi.org/10.2174/1874836801812010389

Tezel, A., Aziz, Z., Koskela, L., & Tzortzopoulos, P. (2016). Visual management condition in highways construction projects in England. 24th Annual Conference of the International Group for Lean Construction, 133–142. https://www.scopus.com/inward/record.uri?eid=2-s2.0-84995932405&partnerID=40&md5=473c5d3bba4c3fc79251bb29bc9ee330

Tezel, A., Koskela, L., & Aziz, Z. (2018). Lean thinking in the highways construction sector: Motivation, implementation and barriers. Production Planning and Control, 29(3), 247–269. https://doi.org/10.1080/09537287.2017.1412522

Tezel, A., & Nielsen, Y. (2013). Lean Construction Conformance among Construction Contractors in Turkey. Journal of Management in Engineering, 29(3), 236–250. https://doi.org/10.1061/(ASCE)ME.1943-5479.0000145

Torp, O., Knudsen, J. B., & Rønneberg, I. (2018). Factors affecting implementation of lean construction. 26th Annual Conference of the International Group for Lean Construction, 1261–1271. https://doi.org/10.24928/2018/0234

Upadhye, N., Deshmukh, S. G., & Garg, S. (2016). Lean manufacturing system implementation barriers: An interpretive structural modelling approach. International Journal of Lean Enterprise Research, 2(1), 46. https://doi.org/10.1504/IJLER.2016.078232

Venkatesh, P. K., & Venkatesan, V. (2021). Experiences from the implementation of last planner system (R) in construction project. In Indian Journal of Engineering and Materials Sciences (Vol. 28, Issue 2, pp. 125–141). NATL INST SCIENCE COMMUNICATION-NISCAIR.

Viana, D. D., Mota, B., Formoso, C. T., Echeveste, M., Peixoto, M., & Rodrigues, C. L. (2010). A survey on the last planner system: Impacts and difficulties for implementation in Brazilian companies. 18th Annual Conference of the International Group for Lean Construction, 497–507.

Vignesh, C. (2017). A case study of implementing last planner system in Tiruchirappalli District of Tamil Nadu—India. International Journal of Civil Engineering and Technology, 8(4), 1918–1927.

Wandahl, S. (2014). Lean Construction with or without Lean—Challenges of implementing Lean construction. 22nd Annual Conference of the International Group for Lean Construction, 97–108.

https://doi.org/10.52202/081537-0005

Womack, J. P., & Jones, D. T. (1997). Lean Thinking—Banish Waste and Create Wealth in your Corporation. Journal of the Operational Research Society, 48(11), 1148–1148. https://doi.org/10.1038/sj.jors.2600967

Xing, W., Hao, J. L., Qian, L., Tam, V. W. Y., & Sikora, K. S. (2021). Implementing lean construction techniques and management methods in Chinese projects: A case study in Suzhou, China. Journal of Cleaner Production, 286. https://doi.org/10.1016/j.jclepro.2020.124944

Zaeri, F., Rotimi, J. O. B., Hosseini, M. R., & Cox, J. (2017). Implementation of the LPS using an excel spreadsheet: A case study from the New Zealand construction industry. Construction Innovation, 17(3), 324–339. https://doi.org/10.1108/CI-01-2016-0002

Biography

Amirsadegh Sedaghat is a researcher in construction management with a focus on lean construction, leadership, and emerging technologies. He holds a Master's degree in Project and Construction Management and a Bachelor's degree in Urban Engineering. His research interests include improving efficiency in construction practices, reducing waste, and integrating tools such as BIM, off-site construction, and automation into modern project delivery. He is passionate about bridging the gap between theory and practice to support more effective, sustainable, and technology-driven construction methods.

Hadi Jafarian is a systems optimization researcher with 14+ years of experience, specializing in project management, system improvement, and research on resilience, sustainability, and optimization in critical systems like healthcare supply chains, using Agile, Scrum, and Lean methodologies.

Adel A. Zadeh is a Professor of Project/Construction Management at Northeastern University in Toronto. He is also serving as the Director of the College of Professional Studies in Toronto. His current research focuses on organizational transformation, AI in project and construction management, agile and hybrid project management, sustainable construction, lean construction, and green buildings.

https://doi.org/10.52202/081537-0005

Project Delivery Strategies for Adaptable Buildings: A comparative case study of two Dutch modular projects

Fransiskus Xaverius Prisyafada, Dr. Ad Straub, Dr. Johan Ninan
Department of Civil Engineering and Geoscience
Delft University of Technology (TU Delft) Delft, The Netherlands
fransiskusxaveriusprisyafada@student.tudelft.nl,
a.straub@tudelft.nl, j.ninan@tudelft.nl

Abstract

In the face of climate urgency and rapid urban change, buildings must become more adaptable to extend their lifespan and reduce environmental impact. Yet, adaptability is often undermined by project delivery methods (PDMs) that are not well equipped to manage, plan, and deliver long-term lifecycle requirements, even though in terms of design, it is technically possible, particularly in modular construction. This research examines how PDMs can be restructured to enhance building adaptability. Using a comparative case study of two Dutch modular projects (a campus office expansion and a relocatable housing development) this study applies a Design Structure Matrix (DSM) and RACI framework to analyze the alignment of technical design, process workflows, and stakeholder roles. The analysis reveals a critical socio-technical misalignment: while modular systems provided technical capability, adaptability was consistently compromised by process-related barriers, including delayed technical coordination, missing lifecycle documentation, and unclear post-use accountability. Findings are synthesized into a structured delivery framework that integrates disassembly planning, modular reuse logic, and iterative feedback loops from the project's outset. This research provides a practical approach for project managers to proactively structure delivery for adaptability. It contributes to the field by demonstrating that successful adaptability is not merely a technical feature but an output of a holistic alignment of the project delivery system.

Keywords: Project delivery method, adaptable buildings, modular construction, design structure matrix, lifecycle planning

https://doi.org/10.52202/081537-0006

Introduction

According to the United Nations Environment Programme, (United Nations Environment Programme, 2025), building and construction sector accounted for 34% of total global CO_2 emissions, making it the highest-emitting sector globally. Therefore, it is important to find ways to reduce buildings' carbon footprint. Previous research has shown that sustainability can be achieved by the choice of processes and materials, especially renewable materials such as wood and its composites (Mlote & Budig, 2022). However, apart from the selection of materials, the longevity and maintainability of buildings are also crucial factors to consider (Mlote et al., 2024).

One way to overcome these problems is to ensure that buildings can accommodate the changing demands of society, community resistance, the need for new infrastructure, the pressure of climate change, and functional obsolescence (Ninan et al., 2024), thereby extending their lifespan and reducing the need for demolition and new construction. This will be of great help because cities are growing rapidly, driven by economic, social, and environmental challenges (Manewa et al., 2016), which leads to high demand for a variety of building types and may require frequent reconstruction. Without adaptability, buildings will be left empty, demolished or rebuilt with new structures to accommodate growing demands and changing needs. Furthermore, if this occurs frequently, it will lead to a scarcity of resources and contribute to environmental degradation.

One of the key principles of sustainability is that development today should not sacrifice the capability of future generations to fulfill their own need (World Commission on Environment and Development, 1987). However, most buildings are traditionally designed as static and permanent structures, which limits their capability to respond to changing demands (Askar et al., 2021). As a result, these outdated structures cause high building vacancy rates (building redundancy), which leads to substantial refurbishment or premature demolition. Both of these imply high costs and create large amounts of waste, which are only reused partially or downcycled into lower-quality products (Manewa et al., 2016).

Specifically in the Netherlands, due to the housing shortage on temporarily available land, which is often subject to future zoning changes, national planning authorities have documented that flexible and relocatable housing models (*flexwonen*) are increasingly being deployed to address this issue (Groot, 2022). This trend represents a policy-driven response to address the urban land issue in densely populated areas. They need fast-construction housing, with permanent quality, but have the capability to modify in the future when the permit ends.

These changes often happen in the whole life cycle of a building without being intended when they were initially designed and constructed (Pinder et al., 2013). One of the main problems is that buildings are made in such a way that many changes or alterations can cause the demolition of building parts or even whole structures (Durmisevic & Brouwer, 2022). Therefore, it is important to implement building adaptability to prolong their lifespan and reduce waste and resource consumption, which also aligns with environmental, social, and economic sustainability goals (Gosling et al., 2013).

Building adaptability is the capability of the building to accommodate change over, even in the use phase, without doing major deconstruction, which also supports resource efficiency, sustainability, and long-term value (Geraedts & van der Voordt, 2014). However, delivering an adaptable building is not just a matter of material selection or design; it is also influenced by how construction projects are managed, planned, and delivered.

 https://doi.org/10.52202/081537-0006

The existing Project Delivery Method (PDM) framework is often siloed from life-cycle and actor involvement programming. In the design, construction, and maintenance phases, they focus solely on the immediate function of a specific part of the project. The project delivery selection does not properly consider that the cycles are interconnected (Ahmed & El-Sayegh, 2020). They also do not consider how unmanaged activity interdependence may hinder future building modifications. Moreover, a lack of early stakeholder involvement will limit the potential of building adaptability to maintain or add its value (Pinder et al., 2013). This gap between adaptability and PDM consideration presents a critical opportunity that remains unaddressed.

However, commonly used project management tools, such as PERT, Gantt charts, or the Critical Path Method (CPM), are limited in addressing interdependence and feedback loops (Giancarlo, 2004). These tools primarily model sequential or parallel activities, but not enough to present the dynamic interaction that affects adaptability. To bridge this gap, this research explores a different approach to analyzing project activity interdependence and actor coordination.

The objective of this research is to develop a structured framework for enhancing PDM that supports building adaptability. By utilizing the DSM and RACI matrix. To address this socio-technical challenge, a multi-faceted analytical approach is required. The Design Structure Matrix (DSM) was selected for its proven ability to map and analyze process workflows and information dependencies, making it ideal for visualizing the process system of project delivery (Giancarlo, 2004). To analyze the social system, the RACI matrix was chosen to clarify stakeholder roles and responsibilities, which is critical for understanding coordination and accountability (Lee et al., 2021). By combining these two methods, this research can provide a holistic analysis of the alignment between project activities and the actors responsible for them.

1. Literature Review

1.1 Building Adaptability: Definitions and Classifications

Being adaptable is defined as the capability of being adapted and suitable to or fit with specific situations (Askar et al., 2021). In system engineering, adaptability can be defined as the ability of a system to change to follow changes in its environment (Haberfellner & De Weck, 2005). A primary objective of adaptability is the ability to prolong the proper lifetime of a building (Addis & Schouten, 2004). Most buildings become abandoned before their technical life comes to an end, because there are inevitable mismatches created between the supply of space and the demand for it (Leaman et al., 1998).
The general understanding of adaptability comes from a shift in the conventional perception of buildings as static, finished objects to dynamic systems consisting of objects and processes of construction, change, deconstruction, and reconstruction (Askar et al., 2021). Furthermore, due to the nature of varying terms describing adaptability, summaries of the similarities and differences among the three most recently and commonly used terms to define the adaptability of buildings, that is "Adaptability," "Flexibility," and "Design for Disassembly (DfD)," can be seen in Table 1.

https://doi.org/10.52202/081537-0006

Table 1 Similarities and differences between adaptability, flexibility and DfD (Mlote et al., 2024)

Similarities	Differences			
	Factor	**Adaptable buildings**	**Flexible buildings**	**Design for disassembly (DfD)**
1. All address the evolving needs (allow reconfiguration and repurposing to an extent) 2. All seek to minimize waste - Reduce the need for new construction -Environmentally friendly 3. All aim to maximize the potential use of space	Scope	Long-term view	Short-term changes	End-of-life reuse/recycle
	Cost	Requires upfront investment and planning	Upfront cost-effective but may need frequent maintenance	Upfront investment and planning + Higher labor and time
	Scale	Large-scale projects: on the building scale	Small-scale projects: down to room/floor plan scale	Any/Both due to focus: mostly down to component or connections scale (per installation)
	Objective	Meeting major changing occupancy needs by extending the value of the structure in the long run	Easy and quick spatial reconfiguration when required	Reducing waste by reusing building materials and components

1.2 Project Delivery for Adaptable Buildings

The Project Delivery Method (PDM) is essential for all individuals involved in project management, regardless of the industry. Moreover, the PDMs discussed in this study apply to the construction industry. A PDM describes the relationship and working methods among project participants in the process of transforming the owner's goal into completed facilities (Chen et al., 2022). It affects construction performance, such as schedule, cost, quality, and efficiency (Al Khalil, 2002). The PDM can be viewed as both a contractual structure and a compensation arrangement that owners utilize to obtain completed infrastructures that meet their needs (Mafakheri et al., 2007)

To analyze and compare how different PDMs support adaptability, a standardized list of project activities is required. This list functions as a foundation for the DSM and RACI analyses in this research. Traditional construction project management neglects activities that are critical for long- term adaptation, such as Design for Demolition (DfD), post-occupancy planning, and life-cycle documentation. Therefore, a specific activity was developed to ensure these critical elements are explicitly included in the analysis.

The 15 core activities used in this study, as can be seen in Table 2, were synthesized from a comprehensive review of academic literature on project delivery, modular construction, and building adaptability. Key sources include frameworks from PDM selection (Zhong et al., 2022), DfD principles (Ostapska et al., 2024), and constructability management (Khan, 2016). The activities were then clustered into key project phase (Planning, Design, Procurement, etc.) to provide a structured and comprehensive view of the project lifecycle. The validity of this activity list was established through multiple academic sources and was confirmed during the case study analysis.

Table 2 list of PDM Activities related to building adaptability

Activity Phase	Activity Clustered	No	Activity	Referenced In
Planning	Establish project intent and	1a	Client needs identification	MDPI Framework

https://doi.org/10.52202/081537-0006

Activity Phase	Activity Clustered	No	Activity	Referenced In
	adaptability goals			(2022)
		1b	Feasibility studies	
		1c	Discussion on recycling goals	Khan (2016)
		1d	Review and implementation of past lessons learned	
	Select delivery method and define responsibilities	2a	Composing project delivery strategy	PDM Evolution (2021)
		2b	Define client-contractor-architect roles	
		2c	Selecting architect and contractor	Khan (2016)
	Appoint key design and execution stakeholders	3a	Conducting surveys	
		3b	Working out Construction Schedule	
		3c	Laying out site efficiently	
Design	Design for disassembly and lifecycle reuse (DfD)	4a	Design planning & lifecycle integration	Ostapska et al. (2024)
		4b	Disassembly-oriented detailing	
		4c	Modular detailing & system rationalization	
		4d	Material strategy & circularity	
	Develop spatial-modular design	5a	Foundational site & design requirement analysis	Khan (2016)
		5b	Design development & technical detailing	
		5c	Circularity, materials, and long-term use planning	Ostapska et al. (2024)
	Structural Design for Adaptability	6a	Design structural parts (column, beam, foundation, etc.)	
		6b	Design lightweight structure	
		6c	Over-dimension elements	
	Design building services (MEP)	7a	Design building services (MEP)	Khan (2016)
	Coordinate technical layout and system	8a	Review of design by other team members	
		8b	Clash detection (using BIM)	
		8c	System routing	
	Conduct constructability review and iteration planning	9a	Design coordination between stakeholders	BIM Workflow Mapping (2015)
		9b	Constructability review (via DSM or BIM)	DSM-Constructability (2016)
		9c	prepare deconstruction plan	Ostapska et al. (2024)
		9d	Prepare a documentation plan for adaptable purposes	
Procurement	Procure systems, materials, and subcontractors	10a	Select supplier/vendors	PDM Performance Comparison (2019)
		10b	Procure materials or services	Giancarlo (2004)
Fabrication	Fabricate modules and prepare off-site logistics	11a	Module fabrication	Giancarlo (2004)
		11b	Off-site material integration	
		11c	Logistics scheduling	
Execution	Prepare site and foundations	12a	Pre-execution setup & planning	Giancarlo (2004)
		12b	Site preparation & construction process enablement	
		12c	Monitoring, coordination & site efficiency	
	Deliver and install modules on site	13a	Core construction execution & sequencing	DB vs DBB Analysis (2017)
		13b	Monitoring, safety, and quality management	Khan (2016)
		13c	Innovation, adaptation, and knowledge retention	
	Test systems and perform commissioning	14	Systems testing and commissioning	DSM Planning Projects (2016)
		15a	Handover of documentation and occupancy	PDM Evolution (2021)

Activity Phase	Activity Clustered	No	Activity	Referenced In
Handover & Post-Occupancy	Handover documentation and plan post-occupancy support		certification	
		15b	Facility management setup and operation	PDM Evolution (2021)

1.3 Identified research gaps

Determine the appropriate project delivery method (PDM) is a crucial managerial decision that significantly influences the success of construction projects (Ahmed & El-Sayegh, 2020). The PDM defines roles on the procurement route, sequence of project phases, and sets a framework for organization, roles, and responsibilities, stating that the choice of PDM is often made ad hoc, with insight into how the decision will influence the final project risk allocation (Engebø et al., 2020). This argument is strengthened by, Lædre et al., (Lædre et al., 2006), maintaining that clients continue to select the same method based on previous projects or habit, without considering what suits each project. Although a wide range of PDMs exists, stakeholders often lack a structured PDM to match the delivery method to the project's adaptability needs.

2. Methodology

2.1 Research Design

This research aims to understand how project delivery models can be adapted to enhance the workflow of adaptable buildings. Exploratory research is the most suitable approach to gaining this understanding since this case has not previously been studied in depth, as it can develop initial ideas, identify problems, and provide direction for further research (Swaraj, 2019).
To gain meaningful insights from the cases, it is necessary to consider the holistic levels, rather than decomposing them into separate variables (Yin & Campbell, 2018). Instead, cross-case synthesis should resemble what has been called a 'case-base' rather than a 'variable-base' approach (Byrne, 2009). A synthesis of findings across multiple cases will generate recommendations on how project delivery models can be adapted to support adaptable buildings.
To integrate the DSM and RACI matrix into research design, the research design was made based on Yin & Campbell (2018). The model is shown in Figure 1.

https://doi.org/10.52202/081537-0006

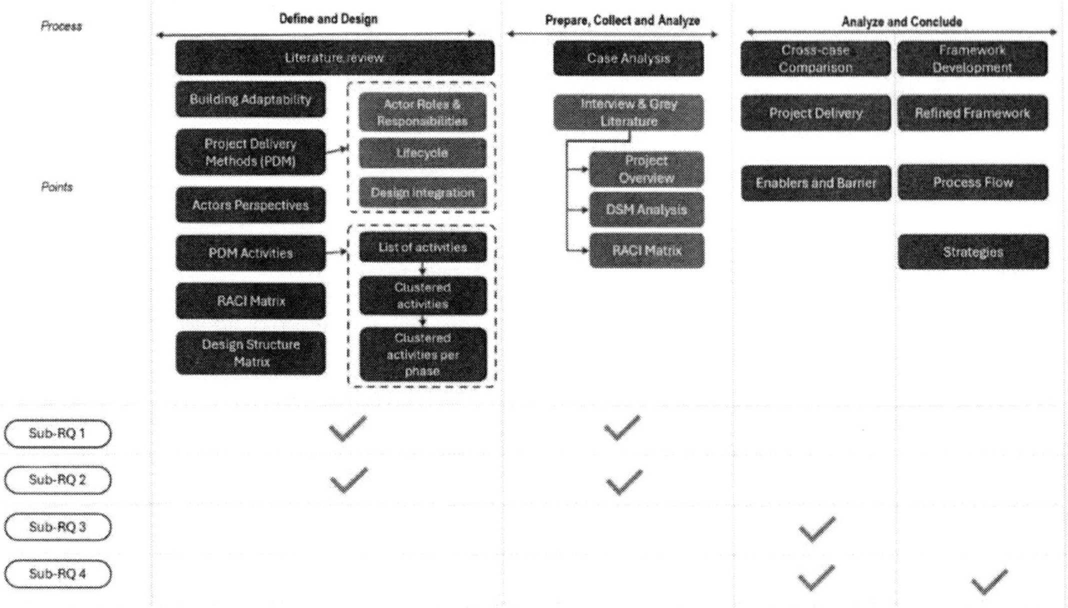

Figure 1 Research Design Framework

This research uses a comparative case study of two Dutch modular projects, which will be referred to as Project A (a campus office expansion) and Project B (a relocatable social housing development) to ensure confidentiality. Case A (the campus office expansion) represents an owner-driven need for flexible and on-site growth. Case B (the relocatable housing development) represents a market- and policy-driven response to pressures on urban land use and the need for temporary housing solutions in densified areas (Groot, 2022). This temporary 15-year permit created a hard deadline for relocation, so it makes lifecycle planning a critical step to take before execution. Comparing these two cases allows for an analysis of how different motivations (internal flexibility versus external policy) shape project delivery strategies for adaptability. The timeline of the project can be seen in Figure 2.

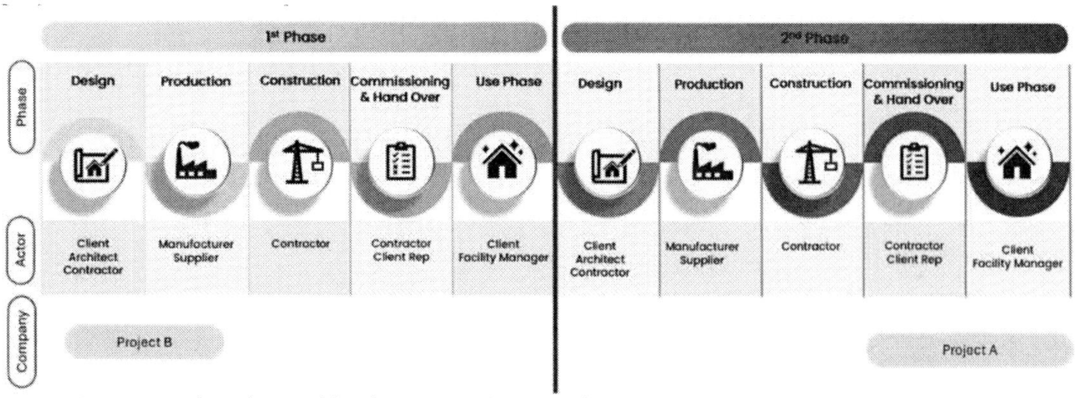

Figure 2 Timeline of project lifecycle: Project A (Campus Lab Expansion) and Project B (Relocatable Housing)

2.2 Definition of Key Actor Roles

To ensure the clarity and consistency throughout the case study analysis, this section defines the key actor roles referred to in this research. The following definition will be used to describe the primary stakeholders and their function within the project delivery process. Client refers to the project's commissioning party, owner, and/or primary end-user. Main Contractor is a single firm that holds the primary contractual responsibility for delivering the entire project, from design coordination through to final execution. The architect or designer is the entity responsible for the building's conceptual, spatial, and aesthetic design. The manufacturer (Modular Manufacturer) is a specialized organization responsible for the off-site, factory-based production of the building's modular components. A specialist contractor is a subcontractor hired for a specific, highly technical scope of work that requires specialized expertise, for example, an MEP contractor, HVAC, or façade.

2.3 Data collection

This research used two primary data collection methods: semi-structured interviews with the project stakeholders and analysis of project documentation. To explore the relationship between project delivery models and building adaptability in practice, semi-structured interviews will be conducted with key stakeholders involved in the case studies, as shown in Table 3.

Table 3 Correspondence Details

Project	Participant number	Stakeholder	Role
Project A	1	Client	Project Coordinator
	2	Client	Supervisor
	3	Client	Program Manager
	4	Main Contractor	Lead Engineer
	5	Main Contractor	Technical Designer
	6	Sub-contractor MEP	Installation Coordinator
	7	Contractor	Project Manager
	8	Contractor	Site Engineer
	9	Contractor	Site Engineer
Project B	10	Client	Location Manager
	11	Client	Junior Project Manager
	12	Contractor	Project Engineer
	13	Contractor	BIM Coordinator
	14	Contractor	Project Manager
	15	Architecture	Architecture

2.4 DSM and RACI as analytical tools in construction

A clear framework to map stakeholders' roles and responsibilities is essential for effectively managing the complex interactions between them. A key tool for this purpose is the responsible assignment matrix, most commonly known as the RACI matrix. The RACI matrix is a straightforward yet powerful tool used in project management to define and communicate the roles and responsibilities of various stakeholders for any given task or activity within a project (Project Management Institute, 2017).

RACI is an acronym that defines four key levels of involvement. Responsible (R): The person or people who do the work to complete the task. They are the "doers" who are responsible for taking

https://doi.org/10.52202/081537-0006

action and implementing it. Accountable (A): The individual who is ultimately answerable for the correct and thorough completion of the task. It is a critical best practice that for any given task, there is only one Accountable person to avoid confusion over ownership. Consulted (C): Stakeholders who are sought out for their input and expertise. This typically involves two-way communication, where their opinions and advice are considered before a decision or action is taken. Informed (I): Individuals who are kept up-to-date on progress or decisions, but do not provide direct input. This is a one-way flow of communication to ensure these stakeholders are aware of project developments.

A Design Structure Matrix (DSM) is defined as a matrix representation of the connections among activities. A DSM provides an idea about various activities of any process that are connected, and what information is needed to start an activity, what activity will be followed by any previous activity, and does task sequencing and iterations (Yassine, 2004). Moreover, Khan (2016) explains that DSM is a representation and analysis tool for system modelling. A DSM displays the relationships between system components in a visual, compact, and analytically advantageous format.

In the context of delivering adaptable buildings, especially where early and continuous collaboration between stakeholders is critical, the RACI matrix serves as a powerful complementary tool to the DSM. While the DSM identifies the technical dependency and information flow between project activities, the RACI matrix clarifies who is responsible for executing those activities and managing the information exchanges.

2.5 Socio-technical Framework in Project Management

To understand the persistent misalignment between technical capability and delivery execution in complex projects, it is useful to adopt a socio-technical perspective. Leavitt's Diamond is a seminal socio-technical framework that provides a lens for this analysis (Leavitt, 1965). The model posits that any organization or project is a system of four interdependent variables: Technology, Tasks, Structure, and People (Actors).

3. Results and Discussion

3.1 Case A: Campus Lab Office Expansion Project

Project A, as can be seen in Figure 3, is located on the TU Delft campus, a living lab for sustainable innovation. In 2017, the Client constructed its original Office Lab 1.0, a modular timber building using CNC-milled Kerto wood modules developed by TU Delft spin-off. In 2024, Project A initiated the expansion of Office Lab 2.0, adding 14 horizontally arranged modules using the same system.

https://doi.org/10.52202/081537-0006

Figure 3 Documentation of Project A

The ambition of Project A is to build an adaptable building was planned from the beginning of the project. However, there is no specific pathway that they follow to control the expansion. The activity list from the previous chapter is used to map the specific activities that the project should do. The organizational chart was pictured in Figure 4.

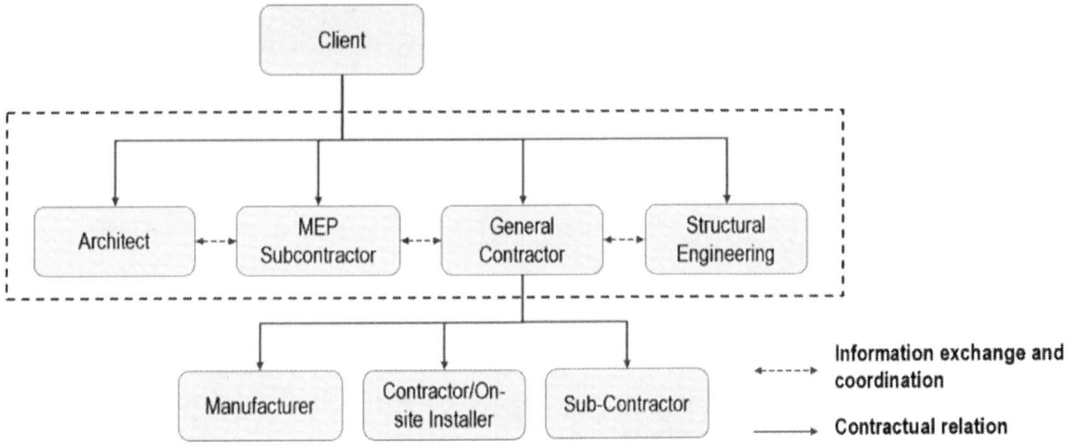

Figure 4 Organizational Chart of Project A

The project's delivery was managed through Design-Bid-Build contracts, where the Client coordinated the design, and the Contractor just executed the design. However, since the modular principle was used in this project, the modular contractor needs to share the limitations of modular with the architect before they start designing. Furthermore, the Modular Contractor or General Contractor runs the project by coordinating the manufacturer and the on-site contractor.

Table 4 show the RACI matrix for every activity that every stakeholder is doing based on the interview and document analysis. The actor-role mapping reveals that adaptability-related outcomes in Project A were not limited by technical capability, but rather by the lack of clearly defined assigned responsibilities across the project lifecycle. This absence of a formal line of collaboration, combined with the late engagement of the technical actor, resulted in project delivery that was able to finish short-term goals (Office Lab 1.0) but underperformed on the adaptability project (Office 2.0).

https://doi.org/10.52202/081537-0006

Table 4 Activity Analysis of Project A

No.	Activity Name	Execution Status	RACI			
			Client	Contractor	Specialized Contractor	Architect
1	Establish project intent and adaptability goals	Modified	R	C	I	I
2	Select the delivery method and define responsibilities	Informal	R	C	I	C
3	Appoint key design and execution stakeholders	Informal	R	C	I	C
4	Develop spatial layout and modular configuration	Fully executed	C	R	I	R
5	Plan the structural system and floor/foundation strategy	Partially addressed	C	R	I	C
6	Design building services (MEP)	Modified	I	C	R	C
7	Coordinate layout and modular constraints	Iterated	I	C	R	C
8	Integrate disassembly and reuse requirements (DfD)	Lacking documentation	A	C	I	I
9	Conduct constructability and feasibility reviews	Minimal	A	I	I	C
10	Procure systems, materials, and subcontractors	Informal procurement	A	C	I	I
11	Fabricate modules and prepare off-site logistics	Fully executed	C	R	I	I
12	Prepare site and foundations	Fully executed	C	R	I	I
13	Deliver and install modules on site	On-site adaptations	C	R	I	C
14	Test systems and perform commissioning	Partial	I	C	R	C
15	Handover documentation and plan post-occupancy support	Missing	A	C	R	C

To understand how coordination and task sequencing influenced adaptability, a DSM was constructed in Table 5. This matrix captures the observed dependencies between activities based on interview data and document analysis.

Table 5 DSM Matrix of Project A

No.	Activity Name		1	2	3	4	5	6	7	8	9	10	11	12	13	14	15
1	Establish project intent and adaptability goals	1		0	0	0	0	0	0	0	0	0	0	0	0	0	0
2	Select delivery method and define responsibilities	2	0		0	0	0	0	0	0	0	0	0	0	0	0	0
3	Appoint key design and execution stakeholders	3	0	0		0	0	0	0	0	0	0	0	0	0	0	0
4	Integrate disassembly and reuse requirements (DfD)	4	1	0	1		0	0	0	0	0	0	0	0	0	0	0
5	Develop spatial layout and modular configuration	5	0	0	0	0		0	0	0	0	0	0	0	0	0	0
6	Structural design for adaptability	6	0	0	0	0	0		0	0	0	0	0	0	0	0	0
7	Design building services (MEP)	7	0	0	0	1	1	0		0	0	0	0	0	0	0	0
8	Coordinate technical layout and system	8	0	0	0	0	0	1	1		0	0	0	0	0	0	0
9	Conduct constructability review and iteration planning	9	0	0	0	0	0	0	0	0		0	0	0	0	0	0
10	Procure systems, materials, and subcontractors	10	0	0	0	0	1	0	1	0	0		0	0	0	0	0
11	Fabricate modules and prepare off-site logistics	11	0	0	0	0	0	0	0	0	0	0		0	0	0	0

https://doi.org/10.52202/081537-0006

No.	Activity Name		1	2	3	4	5	6	7	8	9	10	11	12	13	14	15
12	Prepare site and foundations	12	0	0	0	0	0	0	0	0	0	0	0		0	0	0
13	Deliver and install modules on site	13	0	0	0	0	0	0	0	0	0	0	1	0		0	0
14	Test systems and perform commissioning	14	0	0	0	0	0	0	1	0	0	0	0	0	1		0
15	Handover documentation and plan post-occupancy support	15	0	0	0	0	0	0	0	1	0	0	0	0	0	1	

3.2 Case B: Relocatable Housing

Project B, as shown in Figure 5, is a modular housing project situated in Hilversum, the Netherlands, which was launched to address the region's need for affordable rental housing. The project is built on Diependaalselaan, a site previously used as a temporary event space and will provide approximately 379 relocatable social rental homes. The project also includes public space improvements, communal green areas, and a sustainable energy strategy using all-electric systems. Due to the permit regulation from the municipality of Hilversum, these houses are expected to be relocated elsewhere after 15 years.

Figure 5 Project B Concept drawings

The homes are built using dry-stacked modular units, which are fabricated off-site in the Manufacturer B factory and transported to the site for on-site assembly. The design fulfills the requirements of permanent housing, including acoustic and fire safety regulations, while also allowing for full disassembly and reuse. Even though the superstructure is reusable, the foundation is site-specific and not intended to be relocated.

The Client entrusted the responsibility to build Contractor B, as a Manufacturer and Contractor company, with the prior design by the Architect. Contractor B has to manage off-site manufacturing by assembling modular elements in the factory, such as stairs and MEP systems, from specialized subcontractors. The organizational chart can be seen in Figure 6.

https://doi.org/10.52202/081537-0006

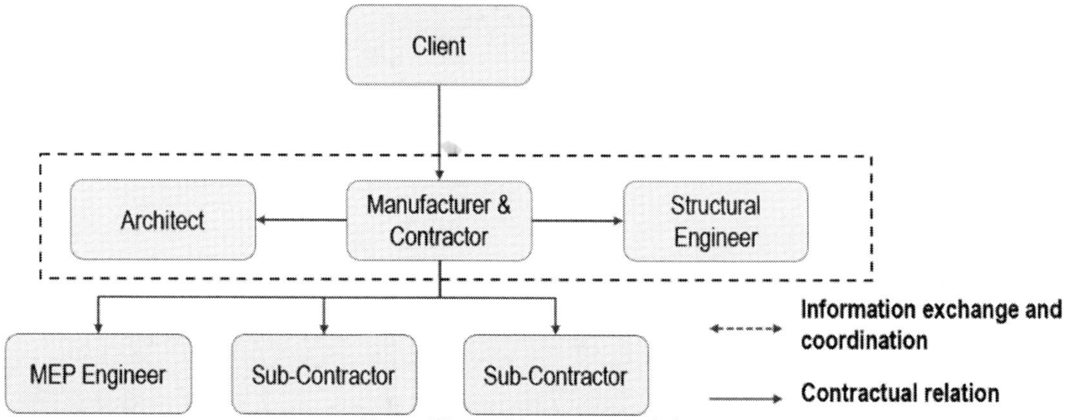

Figure 6 Project B's organizational chart

Project B was delivered using a formal Design & Build (D&B) or turnkey contract. The Client consortium held a contract with the Architect and the Main Contractor (who was also the Manufacturer) but assigned them the responsibility for coordination. Main Contractor has contractual relations with all downstream actors, for example, the MEP Engineer and subcontractors.

3.3 Cross-Case Comparison

4.3.1 Project Delivery Comparison

Interview transcripts were thematically coded by the researcher using the Gioia methodology (Gioia, 2021), starting with open coding to label first-order concepts, as can be seen in Figure 7. These are informant-centric terms or phrases that closely reflect what interviewees said, such as "The ventilation system required more space than anticipated..." was coded as "The existing space for ventilation is not enough to accommodate the new design". Furthermore, these first-order concepts are grouped into broader second-order themes. These are researcher-centric categories that help reveal patterns or theoretical insights, such as "Late Technical Coordination Conflicts". Finally, these themes are synthesized into aggregate dimensions, which represent higher-level constructs that answer the research questions.

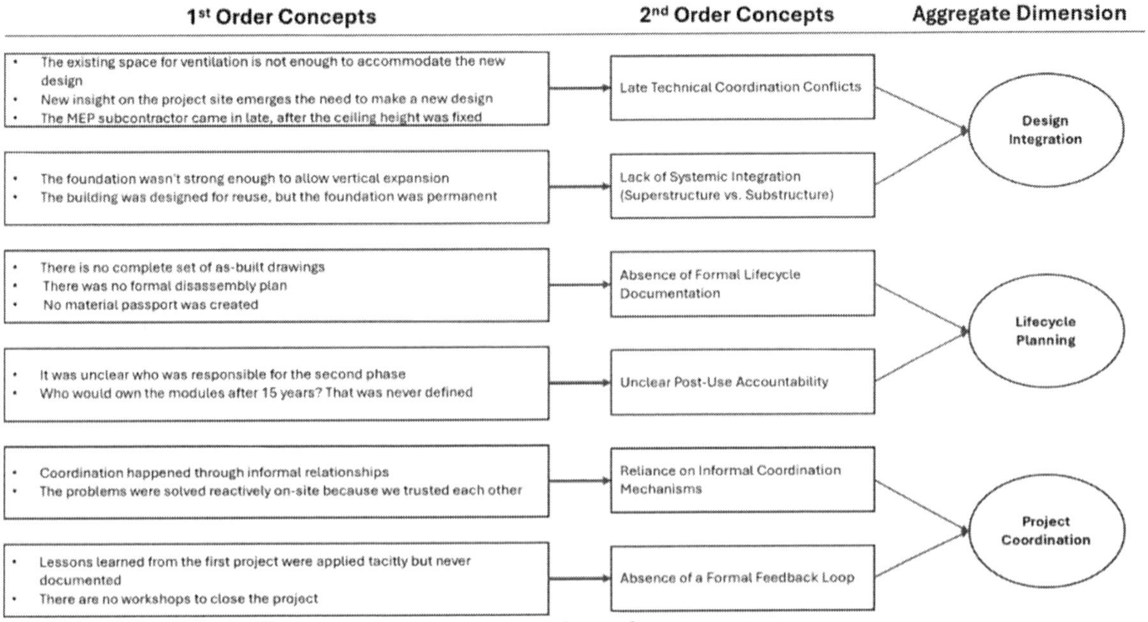

Figure 7 Bottom-up analysis with Gioia

The aggregate dimension is then used to analyse both cases in terms of integration, weaknesses, and lessons, as can be seen in Table 6.

Table 6 Cross-case comparison

Theme	Aspect	Case A - Expansion Office	Case B - Relocatable Housing	Synthesis: Key Lesson for Adaptability
Design Integration	Integration Model	Fragmented, siloed integration, late MEP involvement, linear DSM	Contractor-led D&B model, early alignment of modular constraints and design	Holistic integration is important. Avoid both horizontal silos (lack of cross-disciplinary coordination) Integration between adaptable and non-adaptable components needs to be considered. The scope of integration must encompass all critical systems that impact long-term adaptability.
	Key Weakness	No feedback loops, constructability issues led to rework, MEP routes fixed late	Foundation not designed for relocation	
	Lesson	Horizontal silos blocked systemic adaptability	Future usage of the foundation should be considered	
Lifecycle Planning	Lifecycle Approach	Informal ambition, no formal requirements or deliverables	Clear goal for relocation after 15 years influenced design decisions	Formalized lifecycle governance is imperative. Adaptability must be a contractual deliverable, with clear deliverables (e.g., as-built drawings, DfD manuals), explicit ownership for post-use planning, and a governance structure that persists beyond initial handover.
	Key Weakness	No accountability for lifecycle planning, no reuse documentation	Responsibility for relocation process left undefined; no material passport or guide	
	Lesson	Lifecycle was a hope, not a structured plan	Adaptability goal defined, but execution plan missing	

https://doi.org/10.52202/081537-0006

Theme	Aspect	Case A - Expansion Office	Case B - Relocatable Housing	Synthesis: Key Lesson for Adaptability
Project Coordination	Coordination Style	Informal, trust-based coordination relying on relationships	Formal, hierarchical coordination through single contractor	Combine formal processes and clear hierarchies for contractual definition, documentation, and accountability of lifecycle goals with collaborative communication and trust for early, iterative, and cross-disciplinary dialogue. Embed collaborative routines within a formal contractual framework.
	Key Weakness	Lack of documentation and formal accountability, lost project memory	Efficient but rigid; limited early-stage input from consultants	
	Lesson	Short-term agility, but high risk for long-term adaptability	Clear roles and schedule, but constrained innovation due to centralization	

Design Integration, as a theme, examines the quality and timing of collaboration between key technical disciplines. For adaptable buildings, where the interaction between architectural form, structural systems, and MEP services is crucial for future flexibility, effective integration is paramount. A comparison of Case A and Case B reveals two distinct approaches to design integration, each with its consequences for achieving adaptability goals. Case A demonstrates a fragmented, reactive process, while Case B showcases a more integrated but still incomplete model.

Lifecycle planning encompasses the strategies and processes to manage a building's future, from its ongoing use to its eventual disassembly, relocation, or reconfiguration. While both Case A and Case B were conceived with adaptability in mind, a comparative analysis reveals a significant gap between their lifecycle ambitions and the reality of their planning. This gap demonstrates that without formal documentation and clear accountability, even the most explicit adaptability goals can remain unrealized.

Project coordination governs the flow of information, decision-making authority, and stakeholder relationships that define a project's delivery. The comparison between Case A and Case B reveals a fundamental tension between informal, relationship-based coordination and formal, process-driven coordination. While both models have distinct advantages, they also present significant risks to achieving long-term adaptability.

An ideal model does not force a choice between agility and structure. Instead, it embeds collaborative routines (like early-stage workshops and integrated design reviews) within a formal contractual framework that holds all parties accountable for the project's long-term adaptability. This balanced approach ensures that adaptability is not only designed efficiently but is also managed robustly throughout the building's entire lifecycle.

4.3.2 Enablers and Barriers

Although modular construction offers the ability to be modified, in terms of project delivery, it still requires specific enablers to ensure the entire process runs smoothly. Conversely, mismanagement of project delivery will become a barrier to efficiently modifying the building. Insight into how the enablers and barriers affect the adaptability process can be seen in Figure 8.

https://doi.org/10.52202/081537-0006

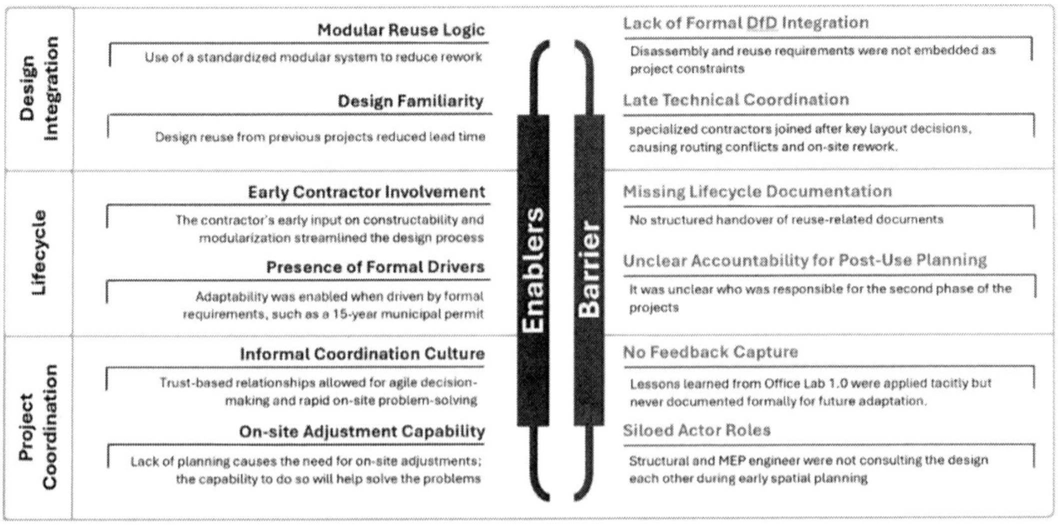

Figure 8 Enablers and Barriers of Implementing Adaptable Buildings

Figure 8 synthesizes the key enablers and barriers to delivering adaptable buildings, as identified from the cross-case analysis. The framework is structured around the three core themes of the research: Design Integration, Lifecycle, and Project Coordination. The left column details the Enablers, which is favorable conditions and practices that were found to support adaptability. These include technical factors, such as using a standardized modular system, and procedural factors, such as early contractor involvement. The right column details the Barriers, which is the critical process gaps and structural shortcomings that undermine adaptability goals. These include issues like a lack of formal Design for Disassembly (DfD) integration and unclear accountability for post-use planning.

This visual comparison highlights a central tension in delivering adaptable buildings: while technical and informal enablers can provide a degree of flexibility, they are often negated by systemic barriers rooted in poor planning and fragmented coordination. For instance, the benefit of an "Informal Coordination Culture" is counteracted by the risk of "No Feedback Capture," and the potential of "Modular Reuse Logic" is lost without formal lifecycle documentation. The diagram illustrates that achieving true, lasting adaptability requires a delivery model that not only leverages its enablers but, more importantly, systematically addresses and mitigates its barriers through structured processes and clear governance.

3.4 Framework Development

4.4.1 The Refined Framework

Table 7 and Table 8 show a synthesized DSM matrix and RACI matrix that combines insights from DSM analysis, RACI mapping, and observation from both case studies. It visualizes how the modular construction workflow for adaptable buildings can be structured across the delivery phase, with specific dependencies between activities, to make the entire process seamless.

https://doi.org/10.52202/081537-0006

Table 7 Refined DSM Matrix

No.	Activity Name		1	2	3	4	5	6	7	8	9	10	11	12	13	14	15
1	Establish project intent and adaptability goals	1		0	0	0	0	0	0	0	0	0	0	0	0	0	0
2	Select delivery method and define responsibilities	2	1		0	0	0	0	0	0	0	0	0	0	0	0	0
3	Appoint key design and execution stakeholders	3	1	0		0	0	0	0	0	0	0	0	0	0	0	0
4	Develop spatial layout and modular configuration	4	1	1	1		0	0	0	0	0	0	0	0	0	0	0
5	Plan structural system and floor/foundation strategy	5	0	0	0	1		0	0	0	0	0	0	0	0	0	0
6	Design building services (MEP)	6	0	0	0	1	1		0	0	0	0	0	0	0	0	0
7	Coordinate layout and modular constraints	7	0	0	0	1	1	1		1	0	0	0	0	0	0	0
8	Integrate disassembly and reuse requirements (DfD)	8	0	0	0	1	1	1	1		1	0	0	0	0	0	0
9	Conduct constructability and feasibility reviews	9	0	0	0	1	0	0	0	1		0	0	0	0	0	0
10	Procure systems, materials, and subcontractors	10	0	0	0	0	0	0	0	0	1		0	0	0	0	0
11	Fabricate modules and prepare off-site logistics	11	0	0	0	0	0	0	0	0	0	1		0	0	0	0
12	Prepare site and foundations	12	0	0	0	0	0	0	0	0	0	0	1		0	0	0
13	Deliver and install modules on site	13	0	0	0	0	0	0	0	0	0	0	1	1		0	0
14	Test systems and perform commissioning	14	0	0	0	0	0	0	0	0	0	0	0	0	1		0
15	Handover documentation and plan post-occupancy support	15	0	0	0	1	0	0	0	1	1	0	1	1	1	1	

Table 8 Refined RACI matrix

No.	Activity Name	Client	Architect	Contractor	Specialized Subcontractor
1	Establish project intent and adaptability goals	**R**	C	I	I
2	Select delivery method and define responsibilities	**R**	C	C	I
3	Appoint key design and execution stakeholders	**R**	C	C	I
4	Develop spatial layout and modular configuration	A	C	**R**	C
5	Plan structural system and floor/foundation strategy	A	**R**	C	C
6	Design building services (MEP)	I	C	C	**R**
7	Coordinate layout and modular constraints	I	C	C	I
8	Integrate disassembly and reuse requirements (DfD)	I	C	**R**	C
9	Conduct constructability and feasibility reviews	I	C	**R**	C
10	Procure systems, materials, and subcontractors	I	I	**R**	C
11	Fabricate modules and prepare off-site logistics	I	I	**R**	I
12	Prepare site and foundations	I	I	**R**	C
13	Deliver and install modules on site	I	I	**R**	I
14	Test systems and perform commissioning	I	I	**R**	I
15	Handover documentation and plan post- occupancy support	**A**	C	**R**	I

Design Integration

The delivery flow consolidates fifteen activities into six interconnected domains (planning, design,

https://doi.org/10.52202/081537-0006

procurement, fabrication, execution, and handover). The DSM matrix highlights the high-dependency pathways, such as how early planning choices (Activities 1–3) directly shape design integration (Activities 4–9), and how disassembly and documentation efforts (Activities 4, 9, and 15) are either dependent or missed depending on the project's governance.

In this framework, delivery logic is no longer linear but reflexive. Design coordination (Activities 8a–8c) and constructability reviews (Activities 9a–9d) are explicitly staged to act as mid-stream evaluative loops rather than post-design correction. This resolves one of the key barriers identified in both cases, which is late technical involvement.

Lifecycle Adaptability

Activities such as "Integrate disassembly and reuse requirements" (Activity 4) and "Prepare a documentation plan for adaptable purposes" (Activity 9d) are emphasized as central to adaptability. These are supported by modular detailing and feedback-enabled loops that span fabrication and post-occupancy planning (Activity 15). Unlike in Project A (where these steps were isolated), the framework positions them as an embedded coordination hub.

By connecting the modular reuse logic (11a–11c) to both procurement sequencing and early MEP routing strategies (7a, 8c), the framework ensures that construction and adaptation are not contradictory. On the contrary, they become co-optimized through spatial prefiguration.

Actor Coordination

The embedded RACI logic further strengthens this framework. Each clustered activity group is aligned with responsibility distribution lessons, for example, early contractor involvement (2a), design coordination (8a), and lifecycle continuity (15). These align with roles and gaps observed in the Project A and Project B projects, particularly the need for clear accountability in post-use planning, technical documentation, and disassembly procedures.

The modular diagram highlights the shift from informal, reactive coordination (Project A) to a more layered, cross-disciplinary integration (Project B). Colored arrows and spatial zones illustrate the activity dependencies and the timing of inter-stakeholder handovers, which will enable project teams to visually map risk, role, and resource needs at each phase.

4.4.2 The Process Flow for Adaptability

The process flow in Figure 9 presents a structured sequence of interdependent activities tailored to support adaptable modular construction projects. The flow is derived from both DSM analysis and cross-case insights, which emphasize how early design decisions, stakeholder coordination, and lifecycle planning interact to enable long-term adaptability. The process aligns activities across time and stakeholders, ensuring that adaptability is not an afterthought but a systemic output of design, delivery, and handover. Feedback loops, multi-actor coordination, and early DfD integration are explicitly structured to enhance the project's capacity for change. For example, activities 8 and 9 become the center of design and execution, ensuring that all designs will be constructible on site when executed. After activities 15, there will be another loop of project life-cycle, which in this case is second project life-cycle.

https://doi.org/10.52202/081537-0006

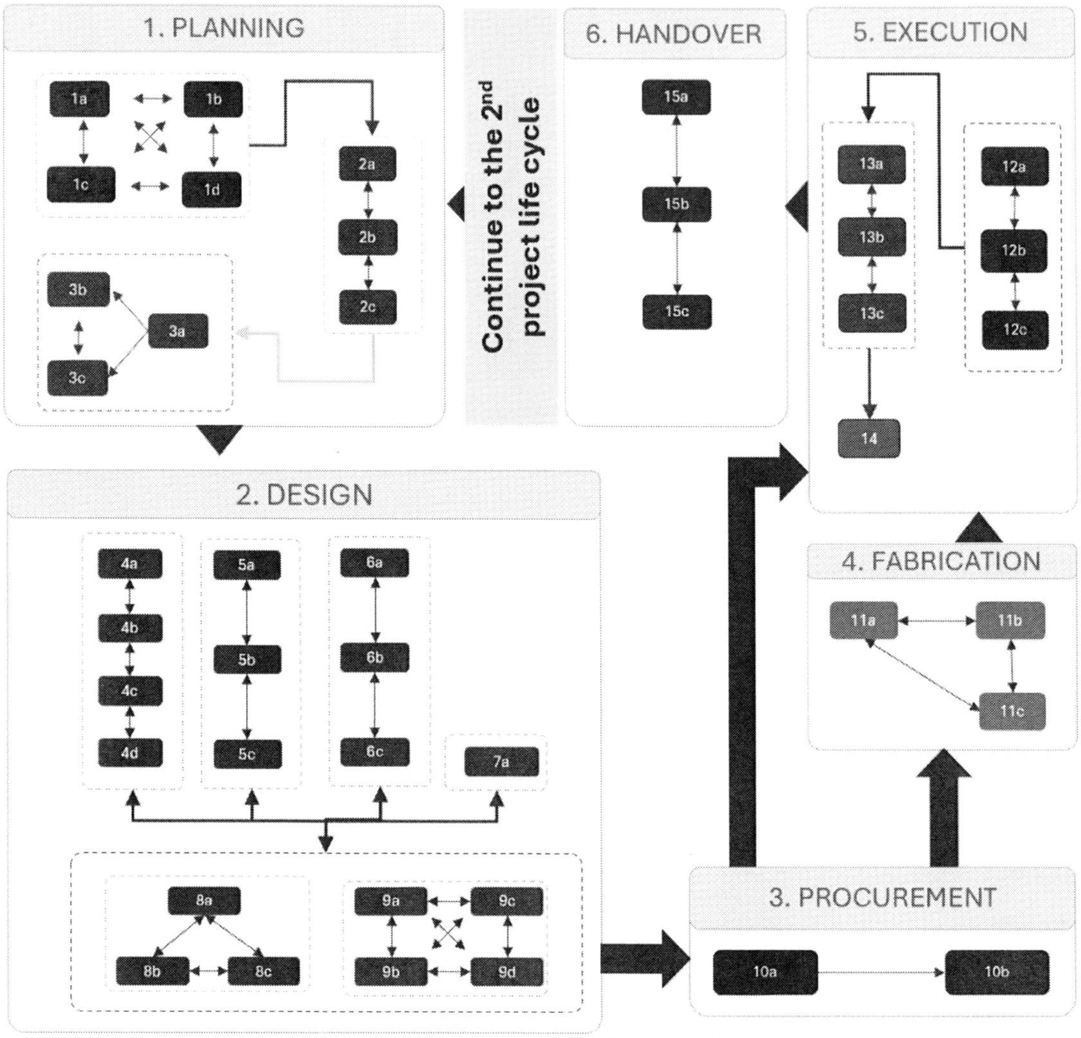

Figure 9 Process Flow of Adaptability Building Project

4.4.3 Core Strategies for Implementation

Figure 10 presents a set of actionable strategies designed to overcome the barriers identified in the case study analysis and systematically embed adaptability into project delivery. The strategies are organized according to the three core themes of this research: Design Integration, Lifecycle, and Coordination, to provide a holistic framework for improvement. Each strategy is a direct response to the shortcomings observed, aiming to shift project delivery from a reactive to a proactive model. The final column, "Related Flow Activity," links each strategic recommendation to specific, numbered activities within the refined project workflow.

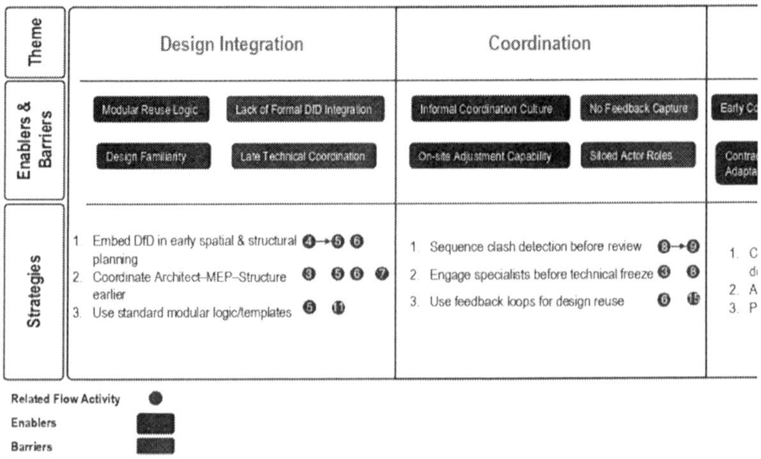

Figure 10 Strategies for Building an Adaptability Project

The content of the strategies reflects a clear focus on formalizing processes and enhancing collaboration at the early stages. Under Design Integration, the strategies concentrate on embedding Design for Disassembly (DfD) principles from the outset and ensuring that architectural, MEP, and structural disciplines are coordinated far earlier than in traditional workflows. Lifecycle strategies address the critical gap in long-term planning by mandating the creation of formal documentation, assigning clear accountability for post-use phases, and requiring DfD plans to be reviewed as a key project deliverable. Finally, the Coordination strategies focus on improving the timing and quality of information flow by sequencing clash detection before constructability reviews and establishing feedback loops for organizational learning and design reuse.

This table serves as a strategic roadmap for project teams. It translates the high-level findings of the research into a coherent set of operational interventions that can be directly implemented in a project plan. By connecting each strategy to specific activities, the table provides a tangible guide for mitigating risks associated with fragmented delivery, ensuring that adaptability is not treated as an optional feature but as a systemic outcome of a well-structured and integrated project delivery process.

4. 5 Discussion

Referring to Leavitt (1965), this research, as can be seen in Figure 11, the Technology component is represented by the advanced modular systems and DfD principles. The Tasks are the 15 core delivery activities required for the project. The Structure is defined by the Project Delivery Method and the process workflows mapped by the DSM. Finally, the People are the project stakeholders, whose roles and responsibilities were analyzed using the RACI matrix.

https://doi.org/10.52202/081537-0006

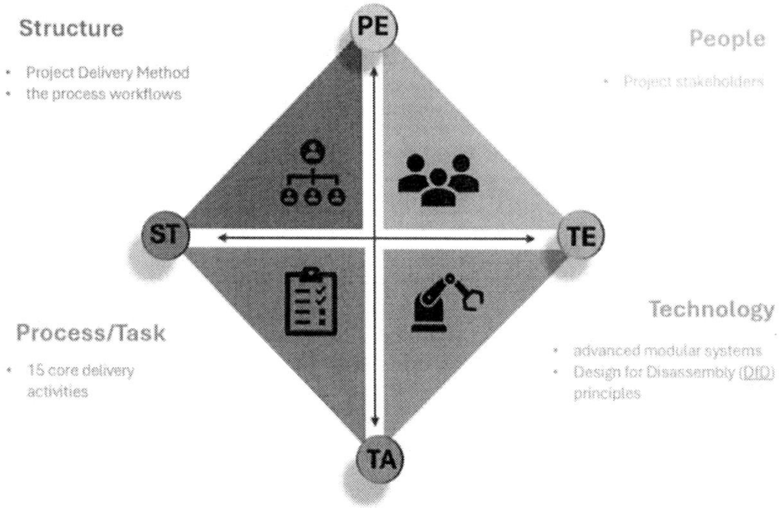

Figure 11 Leavitt's Diamond Diagram

The analysis presented in the preceding chapters reveals a clear and compelling conclusion: the successful delivery of an adaptable building is not merely a technical achievement but a socio-technical one. While the physical design, modular logic, and material selection form the essential technical system that enables change, this research demonstrates that these elements are insufficient on their own. Their potential is often wasted if not supported by a coherent process system (as mapped by the DSM) and a social system that defines stakeholder roles and responsibilities (as analyzed through the RACI matrix). The failures observed in both case studies were not rooted in a lack of technical capability, but rather in the misalignment between these critical systems.

These cases reveal a crucial lesson, visually summarized in Figure 12. This concept illustrates that modularity, on its own, does not guarantee adaptability; it only provides the initial Technical Potential. Realizing this potential is entirely dependent on the strength of the subsequent links in the chain. As the diagram on the left shows, even when the link between a modular system and the project is strong, the overall goal fails if the Delivery Process itself is a "Broken Link." In the case studies, this broken link was represented by siloed workflows, incomplete planning, and a lack of holistic integration, which ultimately severed the connection to a successful Adaptability Outcome.

https://doi.org/10.52202/081537-0006

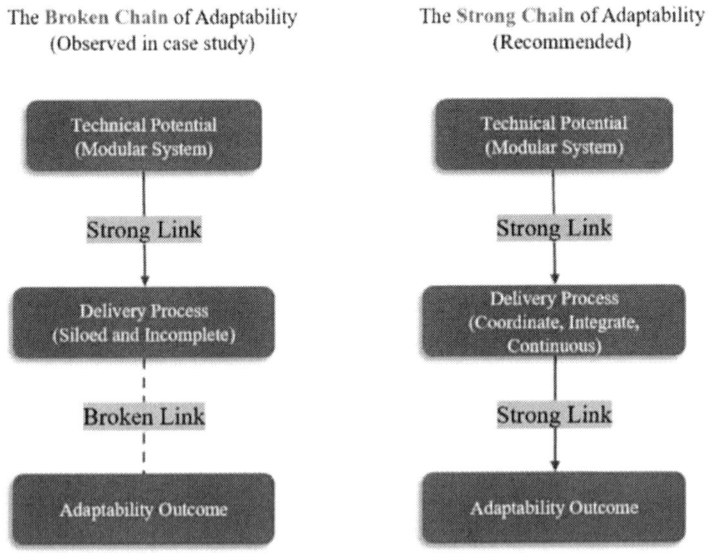

Figure 12 The Chain of Adaptability

The framework proposed in this paper is designed to repair this broken chain. The diagram on the right, "The Strong Chain of Adaptability," illustrates the recommended approach. It posits that an enhanced Delivery Process, one that is coordinated, integrated, and continuous throughout the lifecycle, is the critical connecting link. This includes ensuring that all interconnected systems, from the foundation to the MEP, are part of the adaptability plan, and that the "soft" infrastructure of documentation and lifecycle governance is as strong as the physical modules themselves. Without this comprehensive and holistic delivery framework, even the most advanced modular system risks becoming a fixed asset with unfulfilled promises, its potential lost to a single weak link in the delivery chain.

4 Conclusion

In the face of increasing climate pressure and dynamic urban demands, the construction industry must transition from delivering static objects to creating adaptable and long-life assets. Delivering an adaptable building is not just a matter of material selection or design, but it is also influenced by how construction projects are managed, planned, and delivered. A critical disconnect drove this research: while the need for building adaptability is clear, existing PDMs often fail to properly support it, as they focus on short-term metrics at the expense of long-term value. This study examined how project delivery can be restructured to systematically incorporate adaptability, with a focus on modular and industrialized construction as a key enabler. By applying a socio-technical lens to two Dutch case studies, utilizing the DSM and RACI matrix as analytical tools, this research aimed to develop a more comprehensive delivery framework.

The comparative analysis provides a crucial insight: technical capability alone does not guarantee adaptability. Both case studies possessed the modular systems necessary for physical modification, yet both fell short of their long-term adaptability goals due to significant process and organizational gaps. It fails not because of technical limitations, but because of fragmented workflows, siloed communication between key disciplines, and a lack of formalized lifecycle procedures within the project delivery structure.

https://doi.org/10.52202/081537-0006

In response to these findings, this research has synthesized a structured framework designed to bridge the gap between adaptability ambition and delivery reality. The proposed framework is built upon a foundation of 15 interdependent delivery activities mapped across six project phases. It explicitly integrates critical but often-neglected tasks such as DfD planning, modular reuse logic, and lifecycle feedback loops to ensure adaptability is embedded from the project's inception. By using the DSM to optimize activity sequencing and the RACI matrix to clarify stakeholder roles, the framework provides a clear roadmap for aligning the technical, process, and social systems of a project.

This study offers both practical and theoretical contributions. For project management practitioners, it provides a concrete method to move beyond reactive problem-solving toward a proactive and integrated delivery strategy. It equips them with a structured approach to foster early coordination, establish role clarity, and leverage the full potential of modular systems. This contributes to the field of project management by demonstrating a system approach to aligning project delivery with long-term asset value, which shifts from short-term project completion to long-term lifecycle performance. While the findings are based on two case studies and further research is needed to test the framework in other contexts, this study provides a vital step toward creating a built environment that is not only constructed efficiently today but is also prepared for the challenges of tomorrow.

While this research focuses on the context of adaptable construction, the core findings speak to a more universal challenge: the management of complex systems under long-term uncertainty. The observed 'socio-technical misalignment' is not unique to the building industry but is a common failure mode in fields ranging from software development to public policy, where static project delivery models are often misapplied to dynamic, evolving problems.

https://doi.org/10.52202/081537-0006

References

Addis, W., & Schouten, J. (2004). *Principles of design to facilitate reuse and recycling*. CIRIA.

Ahmed, S., & El-Sayegh, S. (2020). Critical review of the evolution of project delivery methods in the construction industry. *Buildings, 11*(1), 11. https://doi.org/10.3390/buildings11010011

Al Khalil, M. I. (2002). Selecting the appropriate project delivery method using AHP. *International Journal of Project Management, 20*(6), 469–474. https://doi.org/10.1016/S0263-7863(01)00032-1

Askar, R., Bragança, L., & Gervásio, H. (2021). Adaptability of buildings: A critical review on the concept evolution. *[Manuscript in preparation or Unpublished manuscript – clarify if needed]*

Byrne, D. (2009). Case-based methods: Why we need them; what they are; how to do them. In D. Byrne & C. C. Ragin (Eds.), *The SAGE handbook of case-based methods* (pp. 1–10). SAGE Publications, Inc.

Chen, Q., Feng, H., & Garcia De Soto, B. (2022). Revamping construction supply chain processes with circular economy strategies: A systematic literature review. *Journal of Cleaner Production, 335*, 130240. https://doi.org/10.1016/j.jclepro.2021.130240

Durmisevic, E., & Brouwer, J. (2022). *Design aspects of decomposable building structures*. Inhouse Publishing.

Engebø, A., Lædre, O., Young, B., Larssen, P. F., Lohne, J., & Klakegg, O. J. (2020). Collaborative project delivery methods: A scoping review. *Journal of Civil Engineering and Management, 26*(3), 278–303. https://doi.org/10.3846/jcem.2020.12186

Geraedts, R. P., & van der Voordt, T. J. (2014). Developing a tool to measure building adaptability. *Facilities, 32*(9/10), 548–569.

Giancarlo, O. (2004). *Design structure matrix*.

Gosling, J., Sassi, P., Naim, M., & Lark, R. (2013). Adaptable buildings: A systems approach. *Sustainable Cities and Society, 7*, 44–51. https://doi.org/10.1016/j.scs.2012.11.002

Groot, J. (2022). *Flexwoningen als aanvulling op het woningaanbod*. Planbureau voor de Leefomgeving.

Haberfellner, R., & De Weck, O. (2005). Agile system engineering versus agile system engineering. *INCOSE International Symposium, 15*(1), 1449–1465. https://doi.org/10.1002/j.2334-5837.2005.tb00762.x

Khan, S. (2016). Application of design structure matrix (DSM) for managing projects on the basis of constructability. *Civil Engineering and Architecture, 4*(3), 91–111. https://doi.org/10.13189/cea.2016.040302

Lædre, O., Austeng, K., Haugen, T. I., & Klakegg, O. J. (2006). Procurement routes in public building and construction projects. *Journal of Construction Engineering and Management, 132*(7), 689–696. https://doi.org/10.1061/(ASCE)0733-9364(2006)132:7(689)

Leaman, A., Bordass, B., & Cassels, S. (1998). Flexibility and adaptability in buildings: The "killer" variables. *[Clarify publication if available]*

Leavitt, H. J. (1965). Applied organizational change in industry: Structural, technical and humanistic approaches. *[Clarify source if available]*

Lee, W., Lee, S., Jin, C., & Hyun, C. (2021). Development of the RACI model for processes of the closure phase in construction programs. *Sustainability, 13*(4), 1806. https://doi.org/10.3390/su13041806

Mafakheri, F., Dai, L., Slezak, D., & Nasiri, F. (2007). Project delivery system selection under uncertainty: Multicriteria multilevel decision aid model. *Journal of Management in Engineering, 23*(4), 200–206. https://doi.org/10.1061/(ASCE)0742-597X(2007)23:4(200)

Manewa, A., Siriwardena, M., Ross, A., & Madanayake, U. (2016). Adaptable buildings for

https://doi.org/10.52202/081537-0006

sustainable built environment. *Built Environment Project and Asset Management, 6*(2), 139–158. https://doi.org/10.1108/BEPAM-10-2014-0053

Mlote, D. S., & Budig, M. (2022). Load-bearing capacities and pseudo-ductility of carbon fiber-reinforced New Zealand pine timber beams. *Journal of Composites Science, 6*(8), 239. https://doi.org/10.3390/jcs6080239

Mlote, D. S., Budig, M., & Cheah, L. (2024). Adaptability of buildings: A systematic review of current research. *Frontiers in Built Environment, 10*, 1376759. https://doi.org/10.3389/fbuil.2024.1376759

Ninan, J., Lim, Y., De Boer, H., Pesamaa, O., Van Acker, M., Schwab, E., Bernsteiner, J., Soderholm, P., & Hertogh, M. (2024). Embracing multi-functionality in European infrastructure projects: A system of systems perspective. *Sustainable Futures, 8*, 100308. https://doi.org/10.1016/j.sftr.2024.100308

Ostapska, K., Rüther, P., Loli, A., & Gradeci, K. (2024). Design for disassembly: A systematic scoping review and analysis of built structures designed for disassembly. *Sustainable Production and Consumption, 48*, 377–395. https://doi.org/10.1016/j.spc.2024.05.014

Pinder, J., Iii, R. S., & Saker, J. (2013). Stakeholder perspectives on developing more adaptable buildings. *Construction Management and Economics, 31*(5), 440–459. https://doi.org/10.1080/01446193.2013.798007

Project Management Institute. (2017). *A guide to the project management body of knowledge (PMBOK® Guide)* (6th ed.). Project Management Institute.

Swaraj, A. (2019). *Exploratory research: Purpose and process*. United Nations Environment

Programme. (2025). *Not just another brick in the wall: The solutions exist – Scaling them will build on progress and cut emissions fast. Global status report for buildings and construction 2024/2025*. United Nations Environment Programme. https://doi.org/10.59117/20.500.11822/47214

World Commission on Environment and Development. (1987). *Report of the World Commission on Environment and Development: Our common future*. Oxford University Press.

Yassine, A. A. (2004). An introduction to modelling and analyzing complex product development processes using the design structure matrix (DSM) method. *Journal Urbana, 51*(9), 1–17.

Yin, R. K., & Campbell, D. T. (2018). *Case study research and applications: Design and methods* (6th ed.). SAGE Publications, Inc.

Zhong, Q., Tang, H., & Chen, C. (2022). A framework for selecting construction project delivery method using design structure matrix. *[Clarify publication information if available]*

Biography

Fransiskus Xaverius Prisyafada is a master's student in Construction Management and Engineering at Delft University of Technology (TU Delft), specializing in Project and People. He holds a bachelor's degree in civil and environmental engineering from Gadjah Mada University and is a certified professional engineer (Insinyur) in Indonesia. Before joining TU Delft, Fransiskus worked for years in the Indonesian construction industry, including roles as a project cost controller and a site engineer. His research focuses on integrating adaptability into modular construction projects through improved project delivery strategies. He is passionate about sustainable construction and the concept of a circular economy.

Dr. Ad Straub is Associate Professor of Public Organizations and Public Procurement in construction, Department Management in the Built Environment (MBE) at the Faculty of Architecture and the Built Environment. His teaching and research interests are methods and instruments for asset and property management, supply chain integration and collaborative relationships, circular public procurement strategies, and business models for sustainable transformation.

Dr. Johan Ninan is an Assistant Professor in the Faculty of Civil Engineering and Geosciences, Delft University of Technology (TU Delft), the Netherlands. Previously, he was a Post-Doctoral Fellow at University College London (UCL). He is a guest editor of the Project Management Journal (PMJ) and is part of the international editorial board of journals such as the International Journal of Project Management (IJPM) and Construction Management and Economics (CME).

Acknowledgments

The first author gratefully acknowledges the support of the Indonesia Endowment Fund for Education (LPDP) for providing the full scholarship that enabled this research.
The author also extends appreciation to the professionals and project teams involved in the Project A and Project B case studies for generously sharing their time, insights, and project data.

 https://doi.org/10.52202/081537-0006

Applying Transformational Leadership Theory to Graduate Education: A Pedagogical Framework for Enhancing Critical Thinking and Professional Development

Ravi Kalluri
College of Professional Studies
Northeastern University – San Jose
r.kalluri@northeastern.edu

Abstract

Transformational leadership is one of the most popular approaches to leadership. It involves focusing on followers' needs and helps them get self-actualized. It motivates the followers to exceed expectations. Research shows that employees prefer managers to perform transformational leadership behaviors such as encouraging creativity, recognizing accomplishments, building trust, and inspiring a collective vision. The findings suggest that this leadership approach brings positive psychological gains for both manager and employee. This presentation discusses the author's personal experiences with applying transformational leadership theory in a NU CPS Project Management classroom of international graduate students. A possible pedagogical approach is presented with results that include a psychologically safe classroom in which students feel empowered and encouraged to freely debate the teacher's viewpoint and explore new ideas. The implementation approach is complicated by the international demographics of the class primarily coming from high power distance cultures where a teacher cannot be questioned. The gender of the student plays a significant role as well as the nature of course delivery. A self-assessment questionnaire is presented for teachers to calibrate their transformational teaching behaviors in the classroom.

Keywords: pedagogy, graduate education, transformational leadership, social exchange, psychological safety

https://doi.org/10.52202/081537-0007

1. Introduction

With the growing popularity of e-learning pedagogies, the social exchange between teacher and students is becoming increasingly transactional. It has long been thought that the responsibility for learning rests solely with the teacher. Lauermann (2014). This is an unfortunate misinterpretation of pedagogy over the centuries since the myriad of pedagogy enacted and/or discussed by Socrates within Plato's dialogues. George (2015). Plato's epoch-making dialogues featured dialectics, elaboration by example, allegories, analogies, imagery, story-telling, recollection, and thoughtful inquiry. Educators nowadays spend time on "the what" of teaching while "the how" to engage, enlighten, and nurture the love of learning and desire to know is often ignored. However, this learning approach fails to reap the benefits of emergent learning that comes from making the learning process a shared goal of the teacher and the taught. This is the transformational leadership approach to teaching a graduate class where the end result is not limited to course learning outcomes but also creative learning experiences. These emergent experiences last longer in the students' psyche because they helped build it. Most of this learning is practical application of the theoretical constructs taught in the course which the students can apply on their jobs after graduation. Transformational leadership theory, when applied to graduate education, emphasizes inspiring and motivating students and faculty towards a shared vision for the program's future. This involves fostering a culture of innovation, intellectual stimulation, and continuous improvement while also valuing individual growth and development. Previous studies have focused on the transformational leadership of school leaders (principal, provost, elementary school teacher). Vermeulen et al. (2020) Howell et al. (2022) Kareem et al. (2023) Alzoraiki et al. (2024). This study examines the transformational leadership capability of a graduate school professor (referred henceforth as "teacher").

The following research question guided this study:

RQ1. Does transformational leadership of the teacher foster emergent learning outcomes in a graduate classroom?

2. Literature Review

The principal asset of an education system is intellectual capital. According to García-Morales et al. (2008), transformational leadership (TL) impacts employee attitudes, intellectual capital, and work behaviors. TL brings ushers an entrepreneurial environment, in which students, faculty, coaches, and counsellors perform better and innovate in the classroom. It provides a solution to the extant conflict in the system and the potential to transform and respond to the evolving needs of the education system. It includes the knowledge and skills of different stakeholders, such as researchers, students, faculties and staff. Teachers' involvement in their work and schools will increase because of transformational leaders' actions in classrooms Park et al. (2016).

Transformational leadership has evolved since the term was first coined in 1978 by James M. Burns. "The transforming leader looks for potential motives in followers, seeks to satisfy higher needs, and engages the full person of the follower. The result of transforming leadership is a relationship of mutual stimulation and elevation that converts followers into leaders and may convert leaders into moral agents" (p. 4). It is important for transformational leaders to develop

https://doi.org/10.52202/081537-0007

those around them for the good of the organization or cause. Bass (1985) extended Burns' definition by expanding the relationship between leaders and followers. Transformational leaders are concerned with trust, respect, and allegiance because when followers exhibit these traits, they are more likely to want to accomplish tasks to help reach the mission and vision set forth by the transformational leader. Transformational leadership has worked well in educational institutions because the leader "seeks to build the organization's capacity to select its purposes and to support the development of changes to practices of teaching and learning" (Hallinger, 2003, p. 330).

According to Bass and Avolio (1994), the four key elements of transformational leadership are idealized influence, inspirational motivation, intellectual stimulation and individualized consideration. An appropriate combination of these factors may improve productivity and efficiency and lead to a knowledge-based economy, managing and developing intellectual capital within the higher education system.

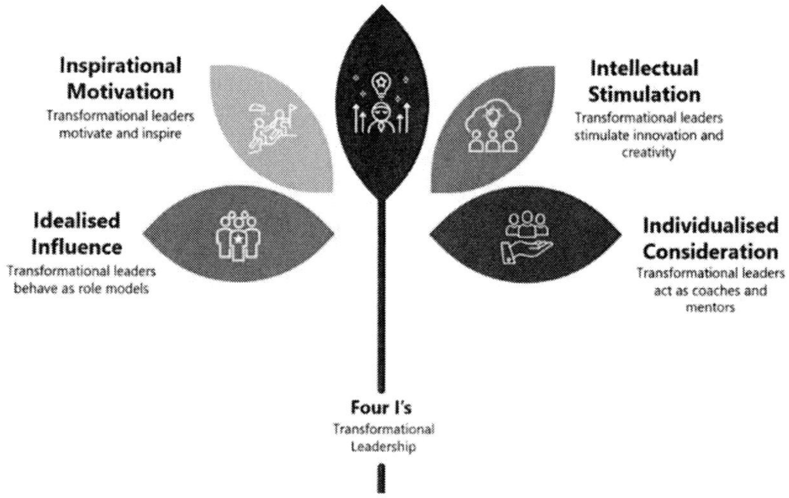

Figure 1. The Four I's of Transformational Leadership Bass (2008)

Transformational leadership is characterized by four key elements, often referred to as the "Four I's": Idealized Influence, Inspirational Motivation, Intellectual Stimulation, and Individualized Consideration. These elements describe how transformational leaders inspire and motivate their teams to achieve a shared vision and improve performance.

Idealized Influence (Charisma): *Transformational leaders act as role models, earning the respect and trust of their followers. They demonstrate high ethical standards and act in ways that inspire others to follow their lead.*
Inspirational Motivation: *Leaders with inspirational motivation communicate a compelling vision and set high expectations, motivating their teams to achieve ambitious goals. They encourage enthusiasm and commitment to the shared vision.*
Intellectual Stimulation: *Transformational leaders challenge the status quo and encourage creativity and innovation within their teams. They foster an environment where new ideas are welcomed and explored.*

Individualized Consideration: *This element focuses on the leader's ability to understand and respond to the unique needs and aspirations of each team member. Leaders provide coaching, mentoring, and support to help individuals develop their potential.*

Kareem et al. (2023) conducted a survey using standardized instruments to measure the leadership style of principals and personal commitment of school teachers in Bengaluru, India. The data was collected from 1,173 school teachers through a questionnaire and analyzed using SPSS V23 statistical software. The study found that transformational leadership had a significant impact on the different types of commitment that teachers possess in school education. The three domains of commitment - commitment towards the institution, student development, and self-development - were positively influenced by transformational leadership.

Transformational leadership has its theoretical foundations in social exchange theory and the social cognitive theory. The social exchange theory by Blau (1964) focuses on the exchange of resources and benefits between individuals in social interactions, where the behavior is influenced by the expectation of recognition and rewards. The social cognitive theory by Bandura (1986) emphasizes the role of cognitive processes, such as observation and imitation, in shaping behavior. Transformational leadership behaviors are therefore likely to promote a teacher's commitment, trust, and gratitude as well as healthy social interaction and group dynamics with students in the classroom.

Transformative leadership has been described by Bass and Avolio (1994) based on the actions of their followers. Such leaders (teachers) are trusted, revered, and admired by their followers (students). Transformational leaders can win people's respect and trust. They change the status quo of the organization through motivational behavior and create a new vision for classroom learning Bakti and Hartono (2022) Lyubykh et al. (2022) Agosto and Roland (2018). According to Firestone and Rosenblum (1988), if teachers supported students' personal growth and upheld a warm and supportive learning environment, the student dropout rate would decline.

Nazari et al. (2012) suggest that transformative leaders have a significant impact on their employees and are successful in enhancing employee commitment. According to Pihie et al. (2011), transformational leaders increase employee commitment by motivating individuals to use innovation and creativity to thrive within the organization. The study suggests that transformational leadership influences teachers' commitment to the school's mission and to the professional learning community in both direct and indirect ways Ross and Gray (2006). Leithwood and Jantzi (2000) found that transformational leadership can lead to higher levels of commitment, ability, and motivation among teachers to develop new approaches to education. In response to transformative leadership, followers have reported similar levels of satisfaction and tendencies to identify such behavior in their leaders Kouzes and Posner (2017) Caza et al. (2021).

Pennell (2023) compared transformational leadership with other leadership theories such as trait-based leadership, emergent leadership, contingency theory, complexity leadership, transactional leadership, and distributed leadership. The author concluded that transformational leadership is a more comprehensive and adaptive approach, particularly within the context of education. According to Assefa and Mujtaba (2025), transformational leaders in education can effectively leverage diversity and technology to create inclusive learning environments, promoting culturally responsive practices and personalized learning experiences that significantly improve student engagement and achievement.

https://doi.org/10.52202/081537-0007

Finally, transformational leadership creates a psychologically safe classroom environment where students feel safe to express themselves with their teacher Myers (2006).

3. Applying Transformational Leadership to Graduate Education

Howell et al. (2022) recommends several activities to instill TL in higher education programs. These include student progress reports on a semester basis and partnerships with organizations outside of the program. Progress reports can provide guidance for the student on their professional goals, networking, and job seeking status. Structured in this manner, the progress reports can provide the student with more personalized feedback than just final course grades. Higher education programs should also create partnerships with organizations outside of their programs but within the university, as this can influence the acquisition of more funding opportunities, more knowledge of contemporary trends and practices, and potentially more impact for the program. Partnerships with organizations outside of the university are also recommended. These can provide opportunities for internships, and other experiential opportunities. This will enable students to gain additional transformational leadership skills and industry connections for employment upon graduation. Professional Associations, NGOs, and others can further expand the learning landscape for students and prepare them for the real challenges of the industry.

Applying the findings from Howell et al. (2022), TL in the graduate classroom can be fostered in the following manner:

Table 1. Transformational Leadership Fostering Activities in Graduate Education

Theme	Key Elements	Description	Example
1. Defining a Compelling Vision	Shared Goals, Inspiration and Motivation	Establish a clear, inspiring vision that aligns with the program's purpose, values, and future aspirations. Teachers should communicate this vision in a way that energizes and motivates students.	A graduate program aiming to lead in [Specific field] research, foster collaboration, and prepare students for impactful careers.
2. Fostering Intellectual Stimulation and Innovation	Encouraging Creativity, Promoting Research	Promote exploration of new ideas, critical thinking, and challenging assumptions. Create a supportive environment for	Organizing workshops on emerging technologies, inviting guest speakers, or supporting novel student research.

https://doi.org/10.52202/081537-0007

		experimentation and innovation.	
3. Cultivating a Culture of Continuous Improvement	Feedback and Reflection, Focus on Growth	Encourage open communication and feedback to identify areas for growth. Emphasize lifelong learning and development for all stakeholders.	Regular program evaluations, feedback sessions, and professional development opportunities.
4. Recognizing and Developing Individual Potential	Individualized Consideration, Mentorship and Guidance	Acknowledge and support the unique strengths and needs of each person. Provide personalized mentorship to help individuals thrive.	Tailored academic advising, connecting student cohorts with mentors, and experiential learning opportunities.
5. Building a Collaborative and Supportive Environment	Teamwork, Open Communication	Foster collaboration and a sense of community among students and faculty. Promote a culture of openness and constructive feedback.	Team-based projects, regular meetings, and encouraging mutual support and transparent dialogue.

By embracing these principles of transformational leadership, graduate programs can create a dynamic and inspiring learning environment that empowers students and faculty to achieve their full potential and contribute to the advancement of their fields.

https://doi.org/10.52202/081537-0007

3. Methodology

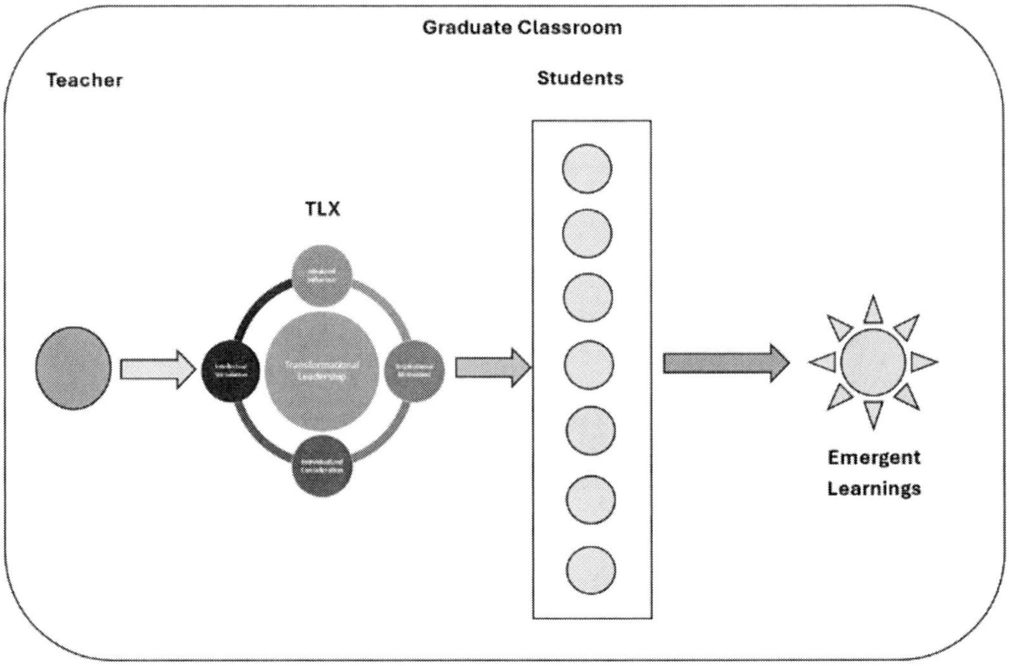

Figure 2. Research Ontology

The methodology used for this study was covert observations made by the author (teacher) over six weeks of PJM 6145 Global Project Management course in Winter 2025.
PJM 6145 is a popular elective course in the MS in Project Management program offered by Northeastern University, College of Professional Studies (CPS). The class had seven international graduate students from India and China. Five of the students were females and two were males. The pedagogical approach was to build a psychologically safe classroom where students feel empowered and encouraged to freely debate the teacher's viewpoint and explore new ideas. All the students had some work experience in their home countries before starting this program. To encourage new ideas, teacher shared his experience as a Project Management Office Director in Seoul, Korea in the first class and asked the students to critique his management approach by writing on the whiteboard. The classes were conducted in a small classroom, on-ground. Given the small class size of seven students, all students were asked to sit in the front row. To promote diversity of ideas, two teams were formed each with students from both countries. The students showed an inclination to sit together their teammates. Every new topic would culminate in a 20-min class dialog period. The goal of this dialog was to critically evaluate the concepts taught by the teacher and come up with counter arguments backed by examples from past experience. Students were encouraged to express verbally or by writing on the whiteboard. In later classes, students proposed sharing their laptop screen with example articles or their notes, with the class.

https://doi.org/10.52202/081537-0007

5. Results and Discussion

Research Question One asks can transformational leadership of the teacher foster emergent learning outcomes in a graduate classroom. In-situ observations were made by the teacher over the six weeks of the course to answer this question. The results are discussed in Table 2.

Table 2. Results of Transformational Leadership Fostering Activities by Week

Week	Topic	TL Tactics	Emergent Learnings *(Outside lecture/readings)*
1.	Foundation & Environmental Factors	Established a clear, inspiring vision for the course with Teacher's Korea expat experience and the critical need for Global PM skills. Teacher asked students to critique his tactics based on past experience. Front row seating of students in 2nd half of class. Two teams formed with students from India and China.	Each student narrated a global assignment experience and compared it to Teacher's Korea role. This uncovered new GPM challenges and strategies. Four students suggested new tactics to manage in the Korea work environment. A student's example of a manager from China HQ in India who made the team work China hours, was powerful. The physical closeness to the teacher's podium improved student confidence to speak.
2.	Cross-Cultural Considerations	Teacher asked Indian students to describe Chinese culture and vice-versa. In addition, we discussed USA, Saudi Arabia, Australia, Sweden, Swiss, France and Brazil's cultures. (Hofstede's Insights)	Students from both countries were surprised to hear some new insights on their countries! Five students discussed their experiences in working with these cultures. We got sensitive insights on Australia and Saudi. French work culture was politely criticized.
3.	Leadership and Governance of Virtual Teams	Students were asked to imagine being sent to a new country to manage a project team	All seven students shared their experience of managing a team in a different country. Students discussed their

https://doi.org/10.52202/081537-0007

		and share their thoughts. Mid-term student feedback was requested.	personal and professional challenges in great detail. Mid-term feedback was positive. Students appreciated the open discussions in class.
4.	Global Teams & Outsourcing, Offshoring & Offsets	Special accommodation (early dismissal) was provided to the Chinese students in class during Chinese New Year. Teacher spent time after the event with the affected students to help them catch-up.	Students from India discussed their experience as a team lead in offshore projects in India, reporting to a foreign project manager. We discussed the communication challenges with outsourced projects in India.
5.	Country Risk Assessment	Teacher asked Indian students to do CRA on China and vice-versa. Students were encouraged to speak openly and give constructive feedback.	We learnt about risks of working in Saudi, Australia and Colombia. We learnt about the country specific risks from the multi-nation Silk Route project of China. Team project submissions, team meetings, and transparent classroom dialog sessions showed positive student engagement.
6. Building a Collaborative and Supportive Environment	Challenges for Global Project Managers Case study group presentations	Teacher cut short his final lecture and asked the students to present their case study. Every team member had to present and reading from the slides was immediately called out by teacher. Both students and teacher asked probing questions and gave constructive feedback to the presenters.	Team presentation content and Q&A displayed a mastery of the subject. The presentations referenced the anecdotes discussed in class and ably related them to the course topics. Teams updated their presentations with the new feedback and submitted again on Canvas that night itself.

The results show that TL tactics used by the teacher in the class room did generate new knowledge and emergent learnings while improving student engagement and performance (RQ1).

A self-assessment questionnaire for teachers to calibrate their transformational teaching behaviors in a graduate classroom is included in the Appendix.

Regarding delimitations, this study was conducted in a single graduate classroom in the Northeastern University CPS, San Jose campus with seven international graduate students in the short period from January 6, 2025 to February 15, 2025.

The study has some limitations as well. The class size was small. The course delivery modality was hybrid (not virtual). There were only two countries (India and China) represented in the class. The class had predominantly female students and there may have been a gender effect.
The teacher is also the observer and may have introduced a self-reporting bias in his observations. There were methodological limitations; using a survey questionnaire might have provided richer data and allowed deeper statistical analysis.
The study results are thus not generalizable to all Northeastern University (NU) CPS classrooms at this time. We need similar data from more hybrid courses, delivery modalities and NU CPS campuses.

6. Conclusion

With the growing popularity of e-learning the role of a teacher has become merely transactional. This is an unfortunate turn of events and it is time to reinstate the pivotal role of a teacher in enabling emergent classroom learning outcomes with transformational leadership. This is a call for all academic institutions to act quickly and with intent before a vital role in the human society disappears.

Acknowledgements

The author would like to thank the students of PJM6145 Global Project Management course section 10 who provided the data for this study. I would also like to thank Dr. Kim Bullington, Adjunct Faculty in the Department of Educational Foundations and Leadership at Old Dominion University for providing the motivation for this enquiry during my doctoral studies.

References

Agosto, V., & Roland, E. (2018). Intersectionality and educational leadership: A critical review. *Review of Research in Education, 42*, 255–285. https://doi.org/10.3102/0091732x18762433

Alzoraiki, M., Ahmad, A. R., Ateeq, A., & Milhem, M. (2024). The role of transformational leadership in enhancing school culture and teaching performance in Yemeni public schools. *Frontiers in Education, 9*, 1413607. https://doi.org/10.3389/feduc.2024.1413607

Assefa, E. A., & Mujtaba, B. G. (2025). Exploring transformational leadership in education by leveraging diversity and technology for inclusive practices. *International Journal of Public Leadership*. https://doi.org/10.1108/ijpl-10-2024-0112

Bakti, R., & Hartono, S. (2022). The influence of transformational leadership and work discipline on the work performance of education service employees. *Multicultural Education, 8*, 109–125.

Bandura, A. (1986). *Social foundations of thought and action: A social cognitive theory*. Prentice Hall.

Bass, B. M. (1985). *Leadership and performance beyond expectations*. The Free Press.

Bass, B. M., & Avolio, B. J. (Eds.). (1994). *Improving organizational effectiveness through transformational leadership*. Sage.

Blau, P. M. (1964). *Exchange and power in social life*. Wiley and Sons.

Burns, J. M. (1978). *Leadership*. Harper & Row.

Caza, A., Caza, B. B., & Posner, B. Z. (2021). Transformational leadership across cultures: Follower perception and satisfaction. *Administrative Sciences, 11*(1), 32. https://doi.org/10.3390/admsci11010032

Firestone, W. A., & Rosenblum, S. (1988). *The alienation and commitment of students and teachers in urban high schools* (ERIC Document No. ED294959). https://eric.ed.gov/?id=ED294959

García-Morales, V. J., Lloréns-Montes, F. J., & Verdú-Jover, A. J. (2008). The effects of transformational leadership on organizational performance through knowledge and innovation. *British Journal of Management, 19*(4), 299–319.

George, L. (2015). Socrates on teaching: Looking back to move education forward. *Procedia – Social and Behavioral Sciences, 174*, 3970–3974.

Hallinger, P. (2003). Leading educational change: Reflections on the practice of instructional and transformational leadership. *Cambridge Journal of Education, 33*(3), 329–352. https://doi.org/10.1080/0305764032000122005

Howell, J. L., Bullington, K. E., Gregory, D. E., Williams, M. E., & Nuckols, W. L. (2022). Transformational leadership in higher education programs. *Journal of Higher Education Policy and Leadership Studies, 3*(1), 51–66. https://doi.org/10.52547/johepal.3.1.51

Kareem, J., Patrick, H. A., Prabakaran, N., B. V., Tantia, V., M. P. M. P. K., & Mukherjee, U. (2023). Transformational educational leaders inspire school educators' commitment. *Frontiers in Education, 8*, 1171513. https://doi.org/10.3389/feduc.2023.1171513

Kouzes, J. M., & Posner, B. Z. (2017). *The leadership challenge: How to make extraordinary things happen in organizations* (6th ed.). John Wiley & Sons.

Lauermann, F. (2014). Teacher responsibility from the teacher's perspective. *International Journal of Educational Research, 65*, 75–89.

Leithwood, K., & Jantzi, D. (2000). The effects of transformational leadership on organizational conditions and student engagement with school. *Journal of Educational Administration, 38*, 112–129. https://doi.org/10.1108/09578230010320064

Lyubykh, Z., Gulseren, D., Turner, N., Barling, J., & Seifert, M. (2022). Shared transformational leadership and safety behaviours of employees, leaders, and teams: A multilevel investigation. *Journal of Occupational and Organizational Psychology, 95*, 431–458. https://doi.org/10.1111/joop.12381

Myers, S. A. (2006). Using leader-member exchange theory to explain students' motives to communicate. *Communication Quarterly, 54*(3), 293–304.

Nazari, K., Zaidatul Akmaliah, L. P., Ramli, B., & Khairuddin, I. (2012). An empirical investigation of lecturers' organizational commitment in technical and vocational colleges in Iran. *Journal of Arts, Science & Commerce, 3*, 1–10.

Park, C. K., Lim, D. H., & Ju, B. (2016). Transformational leadership and teacher engagement in an international context. In *Handbook of Research on Global Issues in Next-Generation Teacher Education* (pp. 22–42).

Pennell, L. (2023). Transformational leadership in education: A comprehensive approach to educational success. *Transformational Leadership in Education*, 1–8.

Pihie, Z. A., Sadeghi, A., & Elias, H. (2011). Analysis of head of departments leadership styles: Implication for improving research university management practices. *Procedia - Social and Behavioral Sciences, 29*, 1081–1090. https://doi.org/10.1016/j.sbspro.2011.11.341

Ross, J. A., & Gray, P. (2006). Transformational leadership and teacher commitment to organizational values: The mediating effects of collective teacher efficacy. *School Effectiveness and School Improvement, 17*, 179–199. https://doi.org/10.1080/09243450600565795

Sunaengsih, C., Komariah, A., Kurniady, D., Suharto, N., Tamam, B., & Julia, J. (2021). Transformational leadership survey. *Mimbar Sekolah Dasar, 8*(1), 41–54. https://doi.org/10.53400/mimbar-sd.v8i1.30468

Vermeulen, M., Kreijns, K., & Evers, A. T. (2020). Transformational leadership, leader–member exchange and school learning climate: Impact on teachers' innovative behaviour in the Netherlands. *Educational Management Administration & Leadership, 50*(3), 491–510. https://doi.org/10.1177/1741143220932582

Appendix

Transformational Leadership Questionnaire for Professors (Self-Assessment)
Adapted from Sunaengsih et al. (2021)

Instructions: Please rate each statement on a scale from 1 (Strongly Disagree) to 5 (Strongly Agree).

1. Idealized Influence
Demonstrating ethical behavior, integrity, and serving as a role model.

1. I act in ways that build trust and respect among students and colleagues.
2. I consistently model the values and behaviors I expect from others.
3. I demonstrate a strong sense of purpose and commitment to academic excellence.
4. I make decisions that reflect fairness and ethical standards.
5. I am seen as a role model by my students and peers.

2. Inspirational Motivation
Communicating a compelling vision and inspiring others to achieve it.

6. I articulate a clear and compelling vision for student success.

https://doi.org/10.52202/081537-0007

7. I inspire students to exceed their own expectations.
8. I express optimism and enthusiasm about the future of learning.
9. I encourage students to see challenges as opportunities for growth.
10. I foster a sense of purpose and meaning in academic work.

3. Intellectual Stimulation

Encouraging innovation, creativity, and critical thinking.

11. I challenge students to think critically and question assumptions.
12. I encourage diverse perspectives and open dialogue in the classroom.
13. I support experimentation with new teaching methods and ideas.
14. I promote problem-solving and inquiry-based learning.
15. I create a learning environment that values curiosity and innovation.

4. Individualized Consideration

Providing personal attention, mentoring, and support.

16. I recognize and respond to the individual needs of my students.
17. I provide mentorship and guidance tailored to each student's goals.
18. I am approachable and available to support students outside of class.
19. I celebrate the unique strengths and contributions of each student.
20. I adapt my teaching to accommodate different learning styles and needs.

Optional Open-Ended Questions

- What specific actions has this professor taken that demonstrate transformational leadership?
- How has this professor influenced your academic or personal growth?
- What suggestions do you have for enhancing their leadership impact?

Biography

Dr. Ravi Kalluri received his Ph.D. in Engineering Management from the Old Dominion University, MBA in Operations and Strategy Management from Northwestern University (Kellogg), M.S. in Electrical Engineering from Stanford University, M.S. in Mechanical Engineering from Ohio State University and B.E. with Honors in Mechanical Engineering from Anna University, Chennai , India. He is presently an Assistant Teaching Professor in the MS Project Management program at Northeastern University, College of Professional Studies. His research centers on student success, faculty development, and education leadership.

https://doi.org/10.52202/081537-0007

AUTHOR INDEX

Northeastern University Global Symposium on Leadership and Project Management
Boston, MA
USA

ISBN 979-8-3313-2541-1

CURRAN ASSOCIATES INC.
proceedings.com

108th Annual Meeting of the South Dakota Academy of Science 2023

Volume 102

Vermillion, South Dakota, USA
23 October 2023

SOUTH DAKOTA ACADEMY OF SCIENCE